TURBO MATHEMATICS

REENA NARANG

Creative CROWS
PUBLISHERS LLP
the publishing & communication people

Published by

Creative CROWS PUBLISHERS LLP
the publishing & communication people

Office: D 73, GF ,Shakarpur, India - 110092 East Delhi, Delhi
Mobiles: +91-9953758404
Emails: ganivpanjrath@yahoo.co.in
Website: creativecrowspublishers.com

Reprint 2024
ISBN-13: 978-81-946870-1-6

ISBN 10: 81-946870-1-2

MRP: Rs. 600/-

Printed at Thomson Press, New Delhi

Dedicated

To Papa and mummy

And

To my darlings, Ashna and Ayesha

Thank You

To the Almighty God, *Waheguru*, for being with me at every step of my journey.

To my parents, for instilling the love of books in me and for believing in me even when I did not.

To my husband, Sudhir, for never disturbing me during the writing of this book.

To my adorable children, Ashna and Ayesha, for being my staunchest supporters.

To all my awesome relatives and friends:

My brothers, Rakesh and Pankaj for giving me the wings to fly and for being gentlest of my critics.

My bhabhis, Rekha and Priya for their continued support and love.

My nephew, Riteek and my nieces, Riddhima, Roopsy, Pooja and Prisha for their trust and encouragement.

My soul sister, Srabani for being the inspiration behind this book and for pushing me beyond my limits.

My fabulous friends, Neeta and Akanksha for years of great friendship, Aprajita and Neha for always having my back, Mayura for constantly motivating me. Divya, for guiding me.

My forever friend Vinita for proof reading and for giving insightful inputs to my work.

My college buddies, Vishaal and Vikas for their unflinching faith in my abilities.

A Special mention:

Without the experience and support from my leaders, peers and the students at Pathways School Noida, this book would not exist.

A big Thank you to the members of my publishing team at Creative Crows Publishing LLP- Ms Tanaaz, Ms. Ganiv and Mr Rehman for their guidance and belief in my work.

Introduction

Mathematics is everywhere in our lives. It is an intrinsic part of nature and of our very existence. It is as much in the rainbows as in the bridges across rivers. The beauty of mathematics lies in its simplicity. Famous mathematician William Paul Thurston once said, "Mathematics is not about numbers, equations, computations or algorithms, it is about UNDERSTANDING".

It is this understanding that a student seeks. We need to understand mathematics as there is no area of life untouched by it. From art to music to business to household management, every sphere needs numerical ability in some form. It is believed that the knowledge of mathematics nurtures the logical, analytical reasoning and problem-solving skills of students. This is the reason why mathematics is an integral part of all competitive examinations like CAT, CLAT, GRE, GMAT, CSAT, SAT. But many students find school math so complicated and the methods so long and confusing that a fear of mathematics starts to set in. And the fear becomes strong enough to start avoiding anything which is remotely numerical. This fear of mathematics can affect anyone, and research has shown that students opt out of certain careers because they require mathematical ability. Even others who are good at mathematics find the conventional methods taught at schools tedious and repetitive. Many adults too fumble the moment they see numbers.

Learning mathematics must be a joyful experience. The aim of this book is to remove the fear of mathematics and make it easy and interesting. It provides step by step approach towards calculations and makes each topic simple to understand. It will help the readers to speed up calculations which are often the stumbling blocks of learning. Along with the speed, it aims to develop the accuracy in calculations as well. This will help in increasing the confidence and self-esteem of students and adults alike. This book will also help students do better in competitive examinations as many of these examinations require quick calculations. This book uses Vedic mathematics sutras as a foundation to these techniques. The process has been simplified and practical examples have been provided in each chapter.

Vedic Mathematics Sutras

Vedic Mathematics is a result of the exhaustive work done by **Late Swami Shri Bharti Krishna Tirathji** (1884 – 1960). He was a great scholar of Sanskrit, English, Mathematics, History and Philosophy. He studied the Atharvaveda, one of the four ancient texts and developed the "*Ganit sutras*" written in it. (Ganit means mathematics and sutras mean formulae). He gave 16 sutras or formulae which cover a wide range of topics of mathematics. These formulae are short, easy to understand and remember. They save a lot of time. They make it possible for a student to do complex calculations in a single line and eliminate the need for lengthy and complex steps. Therefore, Vedic mathematics is considered as a gift of the Vedas to remove the fear or anxiety of mathematics and turn it into a mesmerizing subject.

How to use this book

The chapters in the book have been formulated keeping in the mind the degree of difficulty. The book covers the basic operations in mathematics like subtraction, multiplication, division, square roots, cube roots and many more. The reader will find more than one chapter dedicated to one particular operation. It is, therefore, advised that the reader follows the sequence in which the chapters have been written to ensure that maximum benefit is derived from the book. Practice exercises at the end of each chapter will help in consolidating the concepts done in the chapter.

Hope this book provides you with many "aha" moments !!

Table of Contents

Chapter 1

Base and Complement

1.1 Our Number System

Before we begin our journey of Vedic Math, let us look at our own number system. We follow the decimal number system, or the Hindu-Arabic number system. In this system, we have 10 different digits (0,1,2,3,4,5,6,7,8,9) to represent all the numbers. Also, position of the digit is important. This is the weighted (or positional) number representation, where value of each digit is determined by its position (or their weight) in a number. Position of every digit has a weight which is a power of 10. Each position in the decimal system is 10 times more significant than the previous position, that means numeric value of a decimal number is determined by multiplying each digit of the number by the value of the position in which the digit appears and then adding the products. The least significant position is the unit's place, the position to its right has higher weightage by 10 times and is called ten's place. The next position is hundred's, which is 10 times the ten's place and so on. The positions in the decimal number system are as given below in the table.

The right most digit, 5 in this case, is the least significant and the left most digit, 2 in this case, is the most significant. As you can see, the weight of each digit increases as we move from right to left.

Place value	Hundred Thousand (HTh)	Ten thousand (TTh)	Thousand (Th)	Hundred's (H)	Ten's (T)	Unit (U)
	X 100000	X 10000	X 1000	X 100	X 10	X 1

Example-1 – The number 2135 is interpreted as –

$$2135 = 2 \times 10^3 + 1 \times 10^2 + 3 \times 10^1 + 5 \times 10^0$$

(Note that $10^0 = 1$)

1.2 Base and Complement

In this chapter we will understand the two basic concepts of Vedic Mathematics – Base and Complement.

BASE: Bases are powers of 10. We use powers of 10 for Vedic math as we follow the decimal system of numbers, as explained above. The powers of 10 are as given below:

$$10^0 = 1$$

$$10^1 = 10$$

$$10^2 = 100$$

$$10^3 = 1000$$

$$10^4 = 10000$$

So, the numbers 1,10,100, 1000 .... are called **Bases.**

Base numbers begin with 1 followed by 0's only. Hence, numbers like 11, 110, 200, 1300 are not bases.

Complement: Two numbers are said to be complement of each other if their sum is the next higher base. By this definition, complement of 7 is 3 as 7 + 3 = 10. In other words, 7 and 3 are complements of each other.

Similarly, 35 and 65 are complements of each other as 35 + 65 = 100

This means that the complement of a number can be calculated by subtracting it from the next higher base.

Example 2: (i) Complement of 28

$$100 - 28 = 72$$

(ii) Complement of 892

$$1000 - 892 = 108$$

When we need the bases of larger numbers, we will need to subtract them from larger bases.

Example 3: For complement of 63752, we need to do,

$$100000 - 63752$$

Vedic math has a simplified rule for subtracting from bases, or powers of 10.

1.3 Basic Subtraction Rule:

"All from 9 and the last from 10"

This rule means that, to find the complement of any number, subtract all its digits from 9 and the last or the digit in unit's place from 10. Let us understand this with the help of the previous example.

Example 3: Complement of 63572

Its complement can be found by subtracting 6, 3, 5 and 7 from 9 and subtracting 2 from 10, as shown below:

	9	9	9	9	10
–	6	3	5	7	2
	3	6	4	2	8

So, 36428 is the complement of 63572.

Example 4: Find the complement of 1974

1	9	7	4
(9 – 1)	(9 – 9)	(9 – 7)	(10 – 4)
8	0	2	6

Hence the complement of 1974 is 8026

1.4 Special Cases

Case 1: When 0 is one of the digits in the number but is not at the unit's place

In this case we treat 0 as any other digit and subtract it from 9

Example 5: Find the complement of 6059

6	0	5	9
/	/	\	\
(9 – 6)	(9 – 0)	(9 – 5)	(10 – 9)
3	9	4	1

So, the complement of 6059 is 3941

Case 2: When 0 is at the unit's place in a number

When a 0 comes at the unit's place in a number then while writing the complement, we write the 0 as it is but we treat the digit in the ten's place as the last digit. This implies that we will be subtracting the digit in the ten's place from 10. Let us understand this with the help of an example.

Example 6: Find the complement of 740

7	4	0
/	/	|
(9 – 7)	(10 – 4)	0 (as it is)
2	6	0

Complement of 740 is 260

Case 3: When a number has trailing zeros

If we have a number which has more than one trailing zero, for example, 1300 or 400, while writing the complement, we copy the zeros as it is and treat the first non-zero digit (from the right side or the least significant place) as the last digit. Let us see an example to understand this better.

Example 7: Find the complement of 1300

1	3	0	0
(9 – 1)	(10 – 3)	0	0
8	7	0	0

Complement of 1300 is 8700

Case 4: Complement of a decimal number

In this case, we treat the number like a whole number and follow the rule "All from 9 and the last from 10".

After applying the rule, we place the decimal at the same place as the original number. The following example will clarify this rule.

Example 8: Find the complement of 278.54

2	7	8	.	5	4
(9 – 2)	(9 – 7)	(9 – 8)		(9 – 5)	(10 – 4)
7	2	1	.	4	6

Complement of 278.54 is 721.46.

Similarly complement of 62.38 is 37.62 and complement of 1324.57 is 8675.43

The concepts of base and complement will be used extensively throughout this book.

Summary

Base: Powers of 10 are called bases. Examples 10, 100, 1000, 10000.

Complement: Two numbers are said to be complement of each other if their sum is the next higher base.

Basic rule of subtraction: All from 9 and the last from 10. To find the complement of any number, subtract all its digits from 9 and the last or the digit in unit's place from 10.

Try it yourself

Q1. Find the complement of the following numbers

(a) 643	(b) 726	(c) 845	(d) 1728	(e) 6859
(f) 200	(g) 8098	(h) 70200	(i) 9201	(j) 7200

Q2. Find the complement of the following numbers

(a) 23.42	(b) 58.72	(c) 238.123	(d) 4521.09	(e) 653.215
(f) 1.2346	(g) 28309.1	(h) 1201.05	(i) 63.254	(j) 6403.28

Chapter 2

Subtraction made easy

Subtraction is an arithmetic process of taking one quantity away from another. The result of subtraction is called the **difference**. The number to be subtracted is called **minuend** and the number from which it is being subtracted is called **subtrahend**. For example, if we have

$$9 - 3 = 6$$

Then **9** is the **subtrahend, 3** is the **minuend** and **6** is the **difference.**

In this chapter we are going to learn:

1. Subtraction of any number from the powers of 10 or the bases (10, 100, 1000 and so on)
2. General subtraction

2.1.1 Subtraction of a number from the nearest base

As we studied in chapter 1, the formula for calculating the complement is "All from 9 and the last from 10," the same formula is used to subtract a number from its base.

Example 1: 1000 – 326

It is similar to finding the complement of 326. Using the rule, "All from 9 and the last from 10", we get

3	2	6
↓	↓	↓
(9 – 3)	(9 – 2)	(10 – 6)
6	7	4

Hence, 1000 – 326 = 674

Example 2: 10000 – 4321

Using the complement rule, we get

4	3	2	1
↓	↓	↓	↓
(9 – 4)	(9 – 3)	(9 – 2)	(10 – 1)
5	6	7	9

10000 – 4321 = 5679

Example 3: 1000 – 524

5	2	4
↓	↓	↓
(9 – 5)	(9 – 2)	(10 – 4)
4	7	6

1000 – 524 = 476

2.1.2 Subtraction of a number from a bigger base.

Example 4: 1000 – 27

In this case, we will make the number of digits in the number to be subtracted, 27 in this case, equal to the number of zeros in the base. This can be done by adding 0's in the beginning of the number. Then we can apply the rule *"All from 9 and the last from 10."*

1000 – 27 can be rewritten as 1000 – 027. Now applying the rule,

0	2	7
↓	↓	↓
(9 – 0)	(9 – 2)	(10 – 7)
9	7	3

1000 – 27 = 973

Example 5: 10000 – 273

This can be rewritten as 10000 – 0273 and then the subtraction rule applied to it

10000 – 0273 = 9727

Note: The same method can be applied if we are subtracting a decimal number from a power of 10. It should only be checked that the number of digits before the decimal are equal to the number of 0's in the base.

Example 6: 100 – 26.17

2	6	.	1	7
↓	↓		↓	↓
(9 – 2)	(9 – 6)	.	(9 – 1)	(10 – 7)
7	3	.	8	3

100 – 26.17 = 73.83

Example 7: 1000 – 38. 92

This can be written as 1000 – 038.92 and then solved using the subtraction rule

Hence, 1000 – 38.92 = 961.08

2.1.3 Subtraction of a number from multiples of bases.

The bases are 10, 100, 1000 and so on and their multiples are 20, 300, 5000 and so on. Now we will see how to subtract a number from these multiples.

Example 8: 500 – 27

In this case we will split the multiple of the base into two numbers, one of them being the next higher base to the number to be subtracted.

So, 500 will be written as 400 + 100 as 100 is the next higher base to 27

Now our question becomes 400 + 100 – 27. In this we can do 100 – 27 using the principle *"All from 9 and the last from 10"* and get the answer as 73.

We can then add 73 to 400 to get the final answer as 473.

With practice we can get rid of the intermediate steps and arrive at the answer in a single step.

Example 9: 3000 – 248

3000 – 248

= 2000 + 1000 – 248

= 2000 + 752 (complement of 248)

= 2752

Example 10: 7000 – 43

7000 – 43

= 6000 + 1000 – 43

= 6957

2.1.4 Special Case of subtraction from multiple of powers of 10: when the digits in both the numbers are same

When the number of digits in the subtrahend and the minuend are same then subtract the left most digit and reduce this answer by 1 and then take the complement of the rest of the digits of the minuend. Let us understand this with the help of an example.

Example 11: Find the difference: 8000 – 3246.

As we can see the number of digits in the subtrahend (8000) are same as the number of digits in the minuend, we will follow the following steps:

Step 1: 8 – 3 = 5

Step 2: Reduce the answer in step 1 by 1. So, this is now 4

Step 3: Take the complement of the rest of the digits, 246 in this case. The complement of 246, by applying the rule *"all from 9 and the last from 10"* is 754.

Step 4: The final answer is 4754

Example 12: Find the difference 300 – 123

Step 1: 3 – 1 = 2

Step 2: Reduce the answer in step 1 by 1. So, 2 – 1 = 1

Step 3: Complement of 23 = 77

Step 4: Final answer: 177

2.2 General subtraction without borrowing

We will now understand the rules used in general subtraction sums.

Rule 1: All from 9 and the last from 10

Rule 2: By one more than the one before

We have already seen and used the first rule. Now, let us try to understand the second rule.

The first number of our number system is 0, and using the second rule, one more than 0 is 1.

Similarly, 1 more than 1 is 2 and 1 more than 2 is 3 and so on. This rule is called "*ekadhika*". It has been derived from "ek" meaning 1 and "adhika" meaning more.

Hence, *ekadhika* of 7 is 8; *ekadhika* of 43 is 44;

Ekadhika of a number will be denoted by a dot on the number. For example, $7^{\bullet}$ means *ekadhika* of 7 which is 8. We can also write, $7^{\bullet} = 8$.

Let us now use these two rules to do general subtraction.

Example 13: Find the difference 4326 – 1679

Let us first take a look at the conventional method of subtraction:

	4	$^{1}3$	$^{1}2$	$^{1}6$
-	1	6	7	9
	2	6	4	7

Vedic Method:

Step 1: Starting from the right side, consider the digits at the unit's place. We can see that the number on top is 6 and the number to be subtracted is 9. Since 9 is greater than 6, we will follow the following steps. Take the complement of 9, which is 1, add it to 6. The answer is 7. That is the answer for the unit's place. Also make *ekadhika* of the number at the next place to 9. In this case it is 7. We put a dot on it to indicate *ekadhika*, which means its value is 8.

	4	3	2	6
-	1	6	7•	9
				7

Step 2: In ten's place also, the number to be subtracted is greater. So, we again take the complement of 8 (or 7•) and add it to the number on top. The complement of 8 is 2 and the number on top is 2. Hence the answer is 4. We write this answer in the ten's place and put a dot on the digit in hundred's place and make it *ekadhika*.

	Th	H	T	U
	4	3	2	6
-	1	6•	7•	9
			4	7

Step 3: At the hundred's place, the upper digit is smaller than the lower digit. Hence, we take the complement of the digit in the lower number (6• = 7). In this case the complement is 3 and we add it to the number at the hundred's place in the upper number and get our answer as 6. And we put a dot on the number in the Thousand's place and make it *ekadhika*.

	Th	H	T	U
	4	3	2	6
-	1•	6•	7•	9
		6	4	7

Step 4: In the thousand's place we have 1• = 2. 2 is smaller than 4. Hence, we can directly subtract it and get the answer as 2.

	Th	H	T	U
	4	3	2	6
-	1•	6•	7•	9
	2	6	4	7

Example 14: Find the difference 3895 – 1926

Step 1: In the unit's place we see that the digit to be subtracted is greater than the digit in the unit's place in the upper number. We take the complement of 6, which is 4 and add it to the digit on top and get our answer for the unit's place as 9. We then make the digit in the ten's place as *ekadhika* by putting a dot on it.

	Th	H	T	U
	3	8	9	5
-	1	9	2•	6
				9

Step 2: In the ten's place, the digit in the number to be subtracted is less than the digit in the upper number. Hence, we can directly perform the subtraction (9 - 2•) = 9 – 3 = 6. In this case we DO NOT make the *ekadhika* on the digit in hundred's place.

	Th	H	T	U
	3	8	9	5
-	1	9	2•	6
			6	9

Step 3: In the hundred's place, the digit in the number to be subtracted is greater. Take its complement, 1 in this case and add it to the digit in the upper number (8 + 1) = 9 and make the digit in the thousand's place as *ekadhika* by putting a dot on it.

	Th	H	T	U
	3	8	9	5
-	1•	9	2•	6
		9	6	9

Step 4: In the thousand's place, 3 is greater than 1•. It can be directly subtracted to get 1 as the answer. 3 - 1• = 3 – 2 = 1.

	Th	H	T	U
	3	8	9	5
-	1•	9	2•	6
	1	9	6	9

Example 15: 96247 – 28519

Step 1: Unit's place : Complement of 9 + 7 = 1 + 7 = 8

Ekadhika of ten's place : 1•

	TTh	Th	H	T	U
	9	6	2	4	7
-	2	8	5	1•	9
					8

Step 2: Ten's place: 4 - 1• = 4 – 2 = 2

Ekadhika of hundred's place: not needed.

	TTh	Th	H	T	U
	9	6	2	4	7
-	2	8	5	1•	9
				2	8

Step 3: Hundred's place: Complement of 5 + 2 = 7

Ekadhika of thousand's place

	TTh	Th	H	T	U
	9	6	2	4	7
-	2	8•	5	1•	9
			7	2	8

Step 4: Thousand's place: complement of 8• + 6 = 1 + 6 = 7

Ekadhika of Ten thousand's place

	TTh	Th	H	T	U
	9	6	2	4	7
-	2•	8•	5	1•	9
		7	7	2	8

Step 5: Ten thousand's place: 9 - 2• = 9 – 3 = 6

	TTh	Th	H	T	U
	9	6	2	4	7
-	2•	8•	5	1•	9
	6	7	7	2	8

Final answer: 96247 – 28519 = 67728

Example 16: 4328 – 3129

	Th	H	T	U
	4	3	2	8
-	3	1•	2•	9
	1	1	9	9

Example 17: 43271 – 16779

TTh	Th	H	T	U
4	3	2	7	1
1•	6•	7•	7•	9
2	6	4	9	2

2.2.1 Special Case: *Ekadhika* of 9

We will now consider a case where we have to take the *ekadhika* of 9 in a subtraction problem. According to Vedic math, *ekadhika* of 9 is 10.

Example 18: 2365 – 1799

Step 1: Unit's place: Complement of 9 + 5 = 1 + 5 = 6

Ekadhika of ten's place = 9•

Th	H	T	U
2	3	6	5
1	7	9•	9
			6

Step 2: Ten's place: In the ten's place we have the *ekadhika* of 9 which is 10. Complement of 10 is 0. Hence if we add 0 to the ten's place digit in the upper number we get: 6 + 0 = 6

And we do *ekadhika* of hundred's place

Th	H	T	U
2	3	6	5
1	7•	9•	9
		6	6

Step 3: Hundred's place: Complement of 7• + 3 = Complement of 8 + 3 = 2 + 3 = 5

Ekadhika of Thousand's place

Th	H	T	U
2	3	6	5
1•	7•	9•	9
	5	6	6

Step 4: Thousand's place: Direct subtraction 2 - 1• = 2 – 2 = 0

Th	H	T	U
2	3	6	5
1•	7•	9•	9
0	5	6	6

Final answer: 2365 – 1799 = 566

As we can see, subtraction in Vedic math completely removes the problem of borrowing. It turns a subtraction problem into an addition problem.

Summary:

Subtraction from the Base: All from 9 and the last from 10.

General Subtraction:

1. If the digit in the minuend(lower number) is greater than the digit in the subtrahend(upper number), take the complement of the digit in minuend and add it to the digit of subtrahend AND make the *ekadhika* of the digit in the next place value.
2. If the digit in the minuend is smaller than the digit in the subtrahend, perform the normal subtraction and DO NOT make the *ekadhika* of the digit in the next place value.

Try it yourself

Q. Find the difference in each of the following:

a) 3476 – 1298

b) 4512 – 3459

c) 7861 – 3489

d) 23481 – 13678

e) 97631 – 24981

f) 4673 – 1348

g) 36784 – 12989

h) 45987 – 34979

i) 3217 – 1985

j) 6574 – 3985

Chapter 3

Multiplication

Magical 9

3.1 Interesting facts about number 9

1. The number 9 is a very interesting number in the number series. It is the highest single digit number. If any natural number is multiplied by 9, and the digits of the answer are repeatedly added until it is just one digit, the sum will be nine.

   Example: 1345 x 9 = 12105

   If we add the digits in the answer, 1 + 2+ 1 + 0 + 5 = 9

   Some more examples:

   121 × 9 = 1089 (1 + 0 + 8 + 9 = 18; 1 + 8 = 9)

   234 × 9 = 2106 (2 + 1 + 0 + 6 = 9)

   578329 × 9 = 5204961 (5 + 2 + 0 + 4 + 9 + 6 + 1 = 27; 2 + 7 = 9)

   89765 x 9 = 807885(8 + 0 + 7 + 8 + 8 + 5 = 36; 3 + 6 = 9)

   The only exception to this is multiplying 9 with 0

   9 x 0 = 0 (not equal to 9)

2. Another interesting mathematical fact about the number 9 is that when you add any number to 9, the sum of the digits of the answer is equal to the sum of the digits of the number added to 9.

   9 + 17 = 26 .

   If we add 17 to 9, then the sum of digits in 17 is 1+ 7 = 8 and the answer after addition to 9 is 26. The sum of digits in the answer is 2 + 6 = 8.

   9 + 34 = 43 (3 + 4 = 7; 4 +3 = 7)

   9 + 124 = 133 (1 + 2 + 4 = 7; 1 + 3 + 3 = 7)

3. Take any random number, for example 85763430. Arrange its digits in descending order. You get 87654330. Now arrange its digits in ascending order. You will get 03345678. Find the difference between the two arranged numbers. You will get 84308652. Add the digits of this answer repeatedly till you get a single digit number. It will be 9.

   8+4+3+0+8+6+5+2 = 36; 3 + 6 = 9.

Other examples

321 – 123 = 198 (1 + 9 +8 = 18; 1+8 =9)

4331 – 1334 = 2997 (2+9+9+7 = 27; 2 + 7 = 9)

3.2 Multiplication by 9

In this chapter, we will see that it is easy to multiply any number by a series of 9 or by a number which consists only of 9's

The rules that we will use are:

(i) By one less than one before

(ii) All from 9 and the last from 10

Multiplication: The number to be multiplied is called the **multiplicand**. The number with which we multiply is called the **multiplier**. The result of multiplication is called **product.**

Multiply 789 by 8

789 → Multiplicand

× 8 → Multiplier

6312 → Product

It is important to note here that multiplier and multiplicand can be interchanged. This implies that any one of them can be multiplier and the other one can be multiplicand without affecting the product or the result.

For this chapter we will take the number having 9's as the multiplier.

3.2.1 Multiplicand and the multiplier have same number of digits.

Examples: 45 x 99; 321 x 999; 6754 x 9999

In all these examples the number of digits in the multiplicand and multiplier are equal.

Example 1: Find the product: 6754 x 9999

Conventional method

				6	7	5	4
			X	9	9	9	9
			6	0	7	8	6
		6	0	7	8	6	x
	6	0	7	8	6	x	x
6	0	7	8	6	x	x	x
6	7	5	3	3	2	4	6

Vedic Method

In the Vedic method, the product can be found in a single line in less than 10 seconds.

The answer has two parts, left hand side, referred to as **LHS** and the right-hand side, referred to as **RHS**. The LHS comes from "**by one less than the one before**". This simply means subtract 1 from the multiplicand.

For RHS, we write the complement of the multiplicand, using the rule "**all from 9 and the last from 10**"

Example 2: Find the product 25 x 99

In this case the multiplicand is 25 and the multiplier is 99. Both the numbers have the same number of digits.

We now write the two parts of the answer.

LHS: Subtract 1 from the multiplicand. So, the LHS is **24**

RHS: Complement of multiplicand, which is **75**.

Final answer after combining the LHS and RHS is **2475**

Example 3: Find the product 456 x 999

Multiplicand is 456 and the multiplier is 999

LHS: Subtract 1 from the multiplicand. LHS is **455**.

RHS: Complement of the multiplicand. RHS is **544**.

Final answer after combining the LHS and the RHS is **455544.**

Now let us go back to our first example.

6754 x 9999

				6	7	5	4
X				9	9	9	9
6	7	5	3	3	2	4	6
LHS				RHS			

Example 4: Find the product:

942378 x 999999

LHS: 942377. **RHS**: 057622 Final answer: **942377057622**

3.2.2. Multiplicand has lesser digits than the multiplier

We will now look at examples where the number of digits in the multiplicand is lesser than the number of digits in the multiplier. For example, 45 x 999; 665 x 9999

In this case we will use the same method as described above but before applying the rules, we will make the number of digits same in both the multiplicand and the multiplier. This can be done by adding zero(s) at the beginning of the multiplicand.

Example 5: 45 x 999

In this question, we see that the multiplicand, 45, has two digits and the multiplier, 999, has three digits. We make the digits in the multiplicand equal to the digits in the multiplier by adding a zero at the beginning. We write 45 as 045 and then apply the vedic method to find the product.

LHS: One less than 045 which is 044.

RHS: Complement of 045, which is 955.

Final answer: **044955 or 44955.**

Note: In this case, while calculating the complement of the multiplicand, we assume it to have 3 digits. The complement will thus be obtained from 1000 and not from 100.

Example 6: 665 x 9999

In this case the number of digits in the multiplicand is 3 and the number of digits in the multiplier is 4. We will first make the number of digits equal by adding a zero at the beginning of the multiplicand. The question now becomes 0665 x 9999.

Applying the vedic method we get:

LHS: one less than the multiplicand: 0664

RHS: complement of the multiplicand: 9335

Final answer: **6649335**

NOTE: The LHS and the RHS of the answer have the same number of digits as the number of 9's in the multiplier.

Alternatively, we can also do this multiplication by following the three steps listed below:

Step 1: Reduce the multiplicand by 1

Step 2: Note the number of extra digits in the multiplier. Write that many 9's

Step 3: Write the complement of the multiplicand

Example 7: Find the product 76 x 9999

Step 1: Reducing the multiplicand by 1 we get 76 – 1 = **75**

Step 2: We see that the multiplier has 4 digits whereas the multiplicand has only 2 digits. So we write 2 extra 9's, that is, **99**

Step 3: Complement of 76 is **24.**

Final answer after combining the three steps is **759924**

Example 8: 373 x 999

Step 1: 373 – 1 = 372

Step 2: extra 9 of the multiplier

Step 3: 627

Final answer: **3729627**

3.2.3 Multiplicand has more digits than the multiplier

We will now consider questions where the number of digits in the multiplicand is greater than the number of digits in the multiplier.

Example 9: Find the product 378 x 99

In this case the multiplicand has 3 digits whereas the multiplier has 2 digits.

To get the answer we follow the steps given below:

Step 1: Split the multiplicand into two parts such that the right-hand part has the same number of digits as the multiplier. For this example, we will split 378 as 3|78

Now 3 is the left-hand side and 78 is the right-hand side.

Step 2: Add 1 to the LHS part of the multiplicand. Here, 3 + 1 = 4. Now subtract this number, 4, from the whole multiplicand, that is, 378 – 4 = **374**. This forms the first part of the ANSWER.

Step 3: Complement of the RHS of the split multiplicand will give us the second part of the answer. Complement of 78 is 22. Hence the second part of the answer is **22**.

Combining the first part and the second part of the answer from step 2 and step 3 we get final answer as 37422.

Example 10: 7168 x 99

Step 1: Split the multiplicand. Write it as 71|68. 71 is the LHS and 68 is the RHS. Note that the RHS has the same number of digits as the multiplier.

Step 2: add 1 to LHS. We get 71 + 1 = 72. Subtract 72 from the multiplicand. 7168 – 72 = **7096**

Step 3: complement of RHS= complement of 68 = **32**.

Final answer **709632**.

Example 11: 5678 x 999.

Step 1: 5|678. 5 is the LHS and 678 is the RHS.

Step 2: Add 1 to 5. We get 5 + 1 = 6. Subtract 6 from 5678 we get **5672**.

Step 3: Complement of 678 = **322**

Final answer **5672322**.

Summary:

Case I: Multiplicand and the multiplier have same number of digits: The answer has two parts, left hand side, referred to as LHS and the right-hand side, referred to as RHS. The LHS comes from "by one less than the one before". This simply means subtract 1 from the multiplicand.

For RHS, we write the complement of the multiplicand, using the rule "all from 9 and the last from 10"

Case II: **Multiplicand has lesser digits than the multiplier:** Make the number of digits same in both the multiplicand and the multiplier by adding zero(s) at the beginning of the multiplicand. Then apply the steps of Case I.

Case III: **Multiplicand has more digits than the multiplier:** Split the multiplicand into two parts such that the right-hand part has the same number of digits as the multiplier. Add 1 to the LHS part of the split multiplicand. Now subtract this answer from the whole multiplicand. This forms the first part of the answer. Complement of the RHS of the split multiplicand will give us the second part of the answer. Combine the two parts to get the complete answer.

Try it yourself

Q1. Find the product

(a) 36 x 99 (b) 78 x 99 (c) 69 x 99 (d) 48 x 99 (e) 22 x 99

(f) 345 x 999 (g) 786 x 999 (h) 254 x 999 (i) 861 x 999 (j) 467 x 999

(k) 2345 x 9999 (l) 5674 x 9999

Q2. Find the product

(a) 67 x 999 (b) 465 x 9999 (c) 321 x 9999 (d) 97 x 999 (e) 32 x 9999

(f) 564 x 9999 (g) 221 x 9999 (h) 86 x 999 (i) 94 x 9999 (j) 365 x 99999

Q3. Find the product

(a) 4567 x 999 (b) 468 x 99 (c) 3721 x 99 (d) 497 x 99 (e) 632 x 99

(f) 5164 x 999 (g) 4221 x 999 (h) 886 x 99 (i) 494 x 99 (j) 285 x 99

Chapter 4

Multiplication

Mystical 11

4.1 Interesting facts about number 11

Just like number 9, number 11 is also a mystical number with many interesting facts.

1. 11 is a prime number. It is the smallest two-digit prime number in the decimal number system.
2. If a number is divisible by 11, then the number resulting after reversing the digits, will also be divisible by 11. For example, 72479 is divisible by 11 and 97427 (reversing the digits of 72479) is also divisible by 11.
3. Divisibility by 11: if the difference of the sum of digits at odd places and the sum of its digits at even places, is either 0 or divisible by 11, then the number is divisible by 11.

   Example 1: Is 2547039 divisible by 11?

   (Sum of digits at odd places) – (Sum of digits at even places)

   = (2 + 4 + 0 +9) – (5 + 7 + 3)

   = 15 – 15 = 0

   Hence the number is divisible by 11.

   Example 2: Is 7175753 divisible by 11?

   (Sum of digits at odd places) – (Sum of digits at even places)

   = (7 + 7 + 7 + 3) – (1 + 5+ 5)

   = 24 – 11 = 13

   The number 7175753 is not divisible by 11.
4. Another very interesting test for divisibility by 11 is to separate the number into groups of two consecutive digits (we may add a leading 0 if the number has odd number of digits), and then add up the numbers so formed. If the sum is divisible by 11 then the number is also divisible by 11.

   Example 3: Is 603867 divisible by 11?

   We separate the number into groups of 2 digits: 60, 38 and 67.

   Find their sum: 60+38+67 = 165.

165 is divisible by 11. Hence, 603867 is also divisible by 11.

Example 4: Is 29238 divisible by 11?

Since the number of digits is odd, we add a 0 in the beginning and then separate into groups of two digits and find the sum.

02 + 92 + 38 = 132.

132 is divisible by 11. Hence, 29238 is divisible by 11.

4.2 Multiplication by 11.

In this chapter, we will see that it is easy to multiply any number by a series of 1 or by a number which consists only of 1's.

Example 5: Find the product 436 x 11

For multiplying by a number consisting of 1's we will follow the DOT SANDWICH method. This method comprises the following steps:

Step 1: Sandwich the number by placing dots on both sides of the multiplicand or the number to be multiplied. The number of dots on either side = no. of 1's in the multiplier – 1

In this question the multiplier consists of two 1's. Hence we place one dot on both sides of the number.

Note that the value of each dot is 0 (zero).

• 4 3 6 •

Step 2: Starting from the right dot, keep on adding two digits at a time, until we reach the left-hand side dot.

Step 3: • 4 3 6 • 0 + 6 = 6

(The value of • is 0)

6 is the unit's place digit in the answer.

Step 4: • 4 3 6 • 6 + 3 = 9

9 the ten's place digit in the answer.

Step 5: • 4 3 6 • 4 + 3 = 7

7 is the hundred's place digit in the answer.

Step 6: • 4 3 6 • 4 + 0 = 4

(The value of • is 0)

4 is the thousand's digit in the answer.

Final answer **4796.**

Example 6: Find the product 1371 x 11

Step 1: Sandwich the number by placing one dot on each side of the number

• 1 3 7 1 •

Step 2: Starting from the right dot, keep on adding two digits at a time, until we reach the left-hand side dot.

• 1 3 7 1 • **0 + 1 = 1 → unit's place**

Step 3:

• 1 3 7 1 • **1 + 7 = 8→ ten's place**

Step 4:

• 1 3 7 1 • **3 + 7 = 10 =$_1$0→hundred's place**

Note: since the sum of the two digits in this step is 10, 0 will be written in hundred's place and 1 will carry forward to the next step

Step 5:

• 1 3 7 1 • **3 + 1 + 1 = 5→thousand's place**

(1 from carry forward)

Step 6:

• 1 3 7 1 •**1 + 0 = 1→ ten thousand's place**

Final Answer: 15081

Example 7: Find the product 371 x 111

This time the multiplier has three 1's. Hence, while sandwiching the multiplicand, we will be putting two dots on either side and also we will be taking three digits at a time and finding their sum.

Step 1: Creating a sandwich

• • 3 7 1 • •

Step 2: Starting from the right dot, keep on adding three digits at a time, till we reach the last dot on the left-hand side.

• • 3 7 1 • • $0 + 0 + 1 = 1$

Step 3:

• • 3 7 1 • • $0 + 1 + 7 = 8$

Step 4:

• • 3 7 1 • • $1 + 7 + 3 = 11 = {}_{1}1$

(1 carry forward)

Step 5:

• • 3 7 1 • • $7 + 3 + 0 + 1 = 11 = {}_{1}1$

(1 carry forward)

Step 6:

• • 3 7 1 • • $3 + 0 + 0 + 1 = 4$

Final answer: 371 x 111 = 41181

Example 8: 4256 x 111

Steps:

• • 4 2 5 6 • •

• • 4 2 5 6 • • $\quad$ 0 + 0 + 6 = 6

• • 4 2 5 6 • • $\quad$ 0 + 6 + 5 = 11 = $_{1}1$

• • 4 2 5 6 • • $\quad$ 6 + 5 + 2 + 1 = 14 = $_{1}4$

• • 4 2 5 6 • • $\quad$ 5 + 2 + 4 + 1 = 12 = $_{1}2$

• • 4 2 5 6 • • $\quad$ 2 + 4 + 0 + 1 = 7

• • 4 2 5 6 • • $\quad$ 4 + 0 + 0 = 4

Final answer: 472416

Example 9: 7234 x 111 = $8_{1}0_{1}2974$ = 802974

Example 10: 37562 x 111 = $4_{1}1_{1}6_{1}9_{1}382$ = 4169382

Example 11: 2367 x 1111

In this question, the multiplier has four 1's. Hence, while making the sandwich we will be putting three dots on either side and while taking the sum, we will be taking 4 digits at a time.

• • • 2 3 6 7 • • •

• • • 2 3 6 7 • • • $\quad$ 0 + 0 + 0 + 7 = 7

• • • 2 3 6 7 • • • $\quad$ 0 + 0 + 7 + 6 = 13 = $_{1}3$

• • • 2 3 6 7 • • • $\quad$ 0 + 7 + 6 + 3 + 1 = 17 = $_{1}7$

• • • 2 3 6 7 • • • $\quad$ 7 + 6 + 3 + 2 + 1 = 19 = $_{1}9$

• • • 2 3 6 7 • • • $\quad$ 6 + 3 + 2 + 0 + 1 = 12 = $_{1}2$

• • • 2 3 6 7 • • • $\quad$ 3 + 2 + 0 + 0 + 1 = 6

• • • 2 3 6 7 • • • $\quad$ 2 + 0 + 0 + 0 = 2

Final answer: 2367 x 1111 = 2629737

A recap of the steps is given below:

1. Make a dot sandwich of the number to be multiplied by putting dot(s) on either side of the number. The number of dots is always one less than the number of 1's in the multiplier. If the multiplier is 11, one dot is put on either side for making a sandwich, if the multiplier is 111, 2 dots are put on either side and so on.
2. (a) When the multiplier is 11: Starting from the right most dot, keep on adding two digits at a time till we reach the left most dot.

   (b) When the multiplier is 111: Starting from the right most dot, keep on adding three digits at a time till we reach the left most dot.

   (c) When the multiplier is 1111: Starting from the right most dot, keep on adding four digits at a time till we reach the left most dot.

   The number of digits taken at a time for the sum is always equal to the number of 1's in the multiplier.
3. In any step, if the sum comes as a two-digit number, write only the unit's digit and carry forward the ten's unit to the next step.

4.3 Multiplication by multiples of 11

The steps that are used to multiply a number by 11 can be slightly modified to multiply a number by multiples of 11. For example, if we want to multiply a number by 22, we can write 22 as 2 x 11. Then we can multiply the number by 2 and then by 11 using the same steps.

Example 12: Find the product 3458 x 22

We can write this question as 3458 x 2 x 11

3458 x 2 = 6916

So, now our question transforms to 6916 x 11

Using the steps listed above we can solve this question as:

6916 x 11 = $7_1 6_1 076$ = 76076

Example 13: Find the product 456 x 33

456 x 3 x 11 = 1368 x 11 = $15_1 0_1 48$= 15048

4.4 Multiplication by 12 - 19

Using the Vedic math method eliminates the need to memorize the tables beyond 9. For example, in order to multiply a number by 17, we need to know the table of 7 and not 17.

In this section we will see how to multiply any number by 12, 112, 13, 113 and so on.

Multiplication by 12: For this we will use the rule "twice the penultimate and the ultimate".

Let's try to understand this rule through an example.

Example 14: Find the product: 53 x 12

Step 1: make a dot sandwich by placing one dot on either side of the number.

• 5 3 •

Step 2: Starting from the right-side dot, take 2 digits at a time. We get dot and 3. In this we see that the dot is the last or the ultimate digit and 3 is the second last digit or the penultimate. We now use the rule "twice the penultimate and the ultimate".

We get

• 5 3 • **(2 x 3) + 0 = 6 Twice the penultimate (3) + ultimate(0)**

Step 3: We now take the next two digits. We get 3 and 5. Here 3 is the last digit and 5 is the penultimate digit. Again, applying the same rule, we get

• 5 3 • **(2 x 5) + 3 = 13= $_{1}$3. Since it is a 2- digit answer, we carry 1 forward.**

Step 4: Take the next two digits. They are 5 and the dot. Now 5 is the last digit and the dot is the penultimate digit. Applying the same rule, we get:

• 5 3 • **(2 x 0) + 5 + 1(carry forward) = 6**

Final answer: 636

Example 15: 431 x 12

Step 1: • 4 3 1 •

Step 2: • 4 3 1 • (2 x 1) + 0 = 2

Step3: • 4 3 1 • (2 x 3) + 1 = 7

Step 4: • 4 3 1 • (2 x 4) + 3 = 11= $_1$1

Step 5: • 4 3 1 • (2 x 0) + 4 + 1 = 5

Final answer: 5172

Example 16: Find the product 3451 x 12

Step 1: • 5 1 7 2 •

Step 2: • 5 1 7 2 • (2 x 2) + 0 = 4

Step 3: • 5 1 7 2 • (2 x 7) + 2 = 16 = $_1$6

Step 4: • 5 1 7 2 • (2 x 1) + 7 + 1 = 10=$_1$0

Step 5: • 5 1 7 2 • (2 x 5) + 1 + 1= 12= $_1$2

Step 6: • 5 1 7 2 • (2 x 0) + 5 + 1 = 6

Final answer: 62064

Review for multiplication by 12

Step 1: Make a dot sandwich by putting one dot on either side of the number.

Step 2: Starting from the right-side dot, taking 2 digits at a time, keep on adding twice the second last digit and the last digit. Repeat till you reach the left-side dot.

4.5 Multiplication by 112

This method can be modified to multiply any number by 112, 1112 and so on. For 112, we make a dot sandwich by putting 2 dots on either side of the number and making groups of 3 digits at a time. We apply the rule "Twice the third last digit added to the second last and the last digit".

Example 17: Find the product 543 x 112

Step 1: • • 5 4 2 • •

Step 2: • • 5 4 3 • • (2 x 3) + 0 + 0 = 6 **[3 is the third last digit]**

Step 3: • • 5 4 3 • • (2 x 4) + 3 + 0 = 11 = $_{1}1$

Step 4: • • 5 4 3 • • (2 x 5) + 4 + 3 + 1 = 18= $_{1}8$

Step 5: • • 5 4 3 • • (2 x 0) + 5 + 4 + 1 = 10 = $_{1}0$

Step 6: • • 5 4 3 • • (2 x 0) + 0 + 5 + 1 = 6

Final answer: 60816

Similarly, for multiplying by 1112, we make a dot sandwich by putting three dots on each side of the number and taking 4 digits at a time. We will now double the fourth last digit and add it to the other three digits in the group.

Example 18: Find the product: 6721 x 1112

Step 1: • • • 6 7 2 1 • • •

Step 2: • • • 6 7 2 1 • • • (2 x 1) + 0 + 0 + 0 = 2

Step 3: • • • 6 7 2 1 • • • (2 x 2) + 1 + 0 + 0 = 5

Step 4: • • • 6 7 2 1 • • • (2 x 7) + 2 + 1 + 0 = 17 = $_{1}7$

Step 5: • • • 6 7 2 1 • • • (2 x 6) + 7 + 2 + 1 + 1 = 23 = $_{2}3$

Step 6: • • • 6 7 2 1 • • • (2 x 0) + 6 + 7 + 2 + 2 = 17 = $_{1}7$

Step 7: • • • 6 7 2 1 • • • (2 x 0) + 0 + 6 + 7 + 1 = 14 = $_{1}4$

Step 8: • • • 6 7 2 1 • • • (2 x 0) + 0 + 0 + 6 + 1 = 7

Final answer: 7473752

4.6 Multiplication by numbers from 13 to 19

When we have to multiply any number by 13, 14, 15, 16, 17, 18 or 19, we follow the same rule: We will make a dot sandwich by putting a dot on each side of the number and then we take groups of two digits at a time but this time we will multiply the second last digit by 3, 4, 5, 6, 7, 8, or 9 respectively.

Let us try to understand this with the help of following examples:

Example 19: 478 x 13

Step 1: • 4 7 8 • (3 x 8) + 0 = 24 = $_{2}4$

Step 2: • 4 7 8 • (3 x 7) + 8 + 2 = 31 = $_{3}1$

Step 3: • 4 7 8 • (3 x 4) + 7 + 3 = 22 = $_{2}2$

Step 4: • 4 7 8 • (3 x 0) + 4 + 2 = 6

Answer: 6214

Example 20: Find the product 542 x 17

Step 1: • 5 4 2 •

Step 2: • 5 4 2 • (7 x 2) + 0 = 14 = $_{1}4$

Step 3: • 5 4 2 • (7 x 4) + 2 + 1 = 31 = $_{3}1$

Step 4: • 5 4 2 • (7 x 5) + 4 + 3 = 42 = $_{4}2$

Step 5: • 5 4 2 • (7 x 0) + 5 + 4 = 9

Answer: 9214

Example 21: 671 x 19

Step 1: • 6 7 1 •

Step 2: • 6 7 1 • (9 x 1) + 0 = 9

Step 3: • 6 7 1 • (9 x 7) + 1 = 64 = $_6$4

Step 4: • 6 7 1 • (9 x 6) + 7 + 6 = $_6$7

Step 5: • 6 7 1 • (9 x 0) + 6 + 6 = 12

Answer: 12749

Summary

Multiplication by a series of 1's:

(a) Make a dot sandwich of the number to be multiplied by putting dot(s) on either side of the number. The number of dots is always one less than the number of 1's in the multiplier. If the multiplier is 11, one dot is put on either side for making a sandwich, if the multiplier is 111, 2 dots are put on either side and so on.

(b) Steps for various series of 1's:

a. When the multiplier is 11: Starting from the right most dot, keep on adding two digits at a time till we reach the left most dot.

b. When the multiplier is 111: Starting from the right most dot, keep on adding three digits at a time till we reach the left most dot.

c. When the multiplier is 1111: Starting from the right most dot, keep on adding four digits at a time till we reach the left most dot.

(c) The number of digits taken at a time for the sum is always equal to the number of 1's in the multiplier.

Multiplication by 12:

Step 1: Make a dot sandwich by putting one dot on either side of the number.

Step 2: Starting from the right-side dot, taking 2 digits at a time, keep on adding twice the second last digit and the last digit. Repeat till you reach the left-side dot.

Multiplication by 13 – 19: When we have to multiply any number by 13, 14, 15, 16, 17, 18 or 19, we follow the same rule: We will make a dot sandwich by putting a dot on each side of the number and then we take groups of two digits at a time but this time we will multiply the second last digit by 3, 4, 5, 6, 7, 8, or 9 respectively.

Try it yourself

Q1. Find the product

(a)643 x 11	(b) 5872 x 11
(c) 876 x 11	(d) 65232 x 11
(e) 4321 x 111	(f) 345 x 111
(g) 2567 x 111	(h) 3278 x 111
(i) 7654 x 1111	(j) 7651 x 1111
(k) 9832 x 1111	(l) 654 x 1111

Q2. Find the product

(a)341 x 12	(b) 475 x 12
(c) 389 x 12	(d) 679 x 12
(e) 542 x 12	(f) 2312 x 12
(g) 8731 x 12	(h) 9392 x 12
(i) 876 x 112	(j) 748 x 112
(k) 3781 x 1112	(l) 7421 x 1112

Q3. Find the product

a) 459 x 13	(b) 765 x 15
(c) 1457 x 16	(d) 267 x 18
(e) 2542 x 16	(f) 324 x 13
(g) 832 x 14	(h) 939 x 19
(i) 763 x 17	(j) 418 x 14
(k) 724 x 19	(l) 549 x 18

Chapter 5

Base method of Multiplication

In this chapter, we will see a unique method for multiplying numbers that are near the bases. As studied in the previous chapters, bases are powers of 10. For example, 10, 100, 1000 are al bases. If we have to multiply 998 by 995 using the conventional method, it will take us several steps and time. But using the base method, we can get the answer is a single step in less than 5 seconds.

Selecting a base

We select a particular base for each problem depending on the numbers to be multiplied. If we are asked to multiply 998 by 995, it will be appropriate to take the base as 1000 as both the numbers are very close to 1000. Similarly if we are multiplying 105 by 107, we will choose 100 as the base as both the numbers are close to 100.

Steps to find the answer:

5.1 Both the numbers are below the Base

The answer will have two parts – the left-hand side (LHS) and the right-hand side (RHS). Important point to remember is that the **RHS will have the same number of digits as the zeros in the selected base**. Now we will understand the steps using an example.

Example 1: 97 x 95

Step 1: Select the base and the difference of each of the numbers from the base. In this question, the base is 100 as both the numbers are close to 100. For finding the difference, we see how far these numbers are from the selected base. 97 is -3 from the base. It basically denotes the difference between the number and the base. So it is -3 for 97 and -5 for 95

We write this as

Number	Diff. from the base
97	-3
95	-5

Step 2: Multiply the differences column to get the RHS of your answer. Remember the RHS must have the same number of digits as the base. In this case the RHS will have 2 digits. Multiplication of -3 and -5 yields 15. This has 2 digits, and this forms the RHS of our answer.

	Number	Diff.
	97	-3
x	95	-5
		/ 15

NOTE: If the product of the differences does not give the required number of digits in the RHS, then we will add leading zeros or zeros in the beginning to make the digits equal to the zeros in the selected base.

Step 3: **Cross subtract**: In this will cross subtract first number and difference of the second or second number and the difference of the first number, that is we do (97 – 5) or we do (95 -3), and get the answer as 92. We can do either of the cross difference as they both will yield the same answer. This is the LHS of our answer.

	Number	Diff.
	97	-3
x	95	-5
	92	/ 15

Final answer: 97 x 92 = 9215

Let us solve a few more questions to understand the method better.

Example 2: 994 x 998

Step 1: Selecting a Base and finding the difference.

In this question, both the numbers are near 1000. Hence, the Base selected is 1000. Next we find the difference between the numbers and the base.

994 – 1000 = -6 and 998 – 1000 = -2

	Number	Difference
	994	-6
X	998	-2

Step 2: Finding the RHS.

The RHS in this case will have 3 digits as the Base selected is 1000 which has 3 zeros. But when we find the product of the differences, we get (-6) x (-2)= 12.

We then convert it to a 3-digit number by adding a zero at the beginning. So, our RHS is 012.

	Number	Difference
	994	-6
X	998	-2
	/	**012**

Step 3: Find the LHS.

For the LHS, we do the cross operation, 994 -2 = 992 or 998 – 6 = 992.

Hence, our LHS is 992.

	Number	Difference
	994	-6
X	998	-2
	992 /	012

Final answer: 992012

Example 3: Find the product 9991 x 9989

Step 1: The base will be 10000 and the differences will be -9 and -11 respectively

	Number	Difference
	9991	-9
x	9989	-11

Step 2: Find the RHS.

To find the RHS, we find the product of the differences (-9) x (-11) = 99

Since our base as 4 zeros, we convert this answer into a 4-digit answer by adding zeros in the beginning. So, our RHS is 0099.

	Number	Difference
	9991	-9
x	9989	-11
	/	**0099**

Step 3: Find the LHS

For finding the LHS, we do the cross operation (9991 – 11 = 9980) or (9989 – 9 = 9980). The LHS is 9980.

	Number	**Difference**
	9991	-9
x	9989	-11
	9980 /	0099

Final answer: 99800099

Example 4: Find the product 99993 x 99992

Step 1: Base is 100000

	Number	**Diff.**
	99993	-7
X	99992	-8

Step 2: RHS: (-7) x (-8) = 56

But since the base is 100000, RHS will be written as 00056

	Number	**Diff.**
	99993	-7
X	99992	-8
	/	**00056**

Step 3: LHS: 99993 – 8 = 99985 OR 99992 – 7 = 99985

	Number	**Diff.**
	99993	-7
X	99992	-8
	99985 /	00056

Final answer: 9998500056

Example 5: Find the product: 989 x 980

Base: 1000

	Number	**Difference**	
	989	-11	
X	980	-20	
	969 /	220	RHS (-11 x -20)

Answer: 969220

Special Case: When the digits in the RHS exceeds the number of zeros in the base

Example 6: Find the product: 88 x 91

Base: 100

	Number	Difference
	88	-12
X	91	-9

In this case when we find the RHS by finding the product of the differences, we see that the answer is (-12) x (-9) = 108. But the RHS cannot have more digits than the zeros in the base. Hence, we carry over the 1 in 108 to the LHS

	Number	Difference
	88	-12
X	91	-9
	79 /	108

LHS now becomes 80 and RHS is 08

Answer: 8008

Example 7: Find the product 91 x 85

	Number	Diff.
	91	-9
x	85	-15
	76 /	$_{1}35$

Final answer: 7735

5.2 Both the numbers are above the Base.

Example 8: 105 x 107

The steps are the same.

Step 1: We select the base. Then find the difference between the numbers and the base. But in this case the difference will be positive.

Base: 100

	Number	Difference
	105	+5
x	107	+7

Step 2: RHS: We find the RHS by finding the product of the differences, 7 x 5 = 35. Since the base has 2 zeros, the RHS will also have 2 digits.

	Number	Difference
	105	+5
x	107	+7
	/	35

Step 3: LHS: For finding the LHS, we do the cross operation:
105 + 7 = 112 OR 107 + 5 = 112.

	Number	Difference
	105	+5
x	107	+7
	112 /	35

Final answer: 11235

Example 9: Find the product: 1008 x 1010

Base: 1000

	Number	Difference
	1008	+8
X	1010	+10
	1018 /	080

The RHS will have 3 digits as base is 1000

(the LHS is obtained by cross operation)

Final answer: 1008 x 1010 = 1018080

Example 10: Find the product: 113 x 109

Base: 100

	Number	Difference
	113	+13
x	109	+9
	122 /	$_{1}17$

RHS cannot have more than 2 digits.

Final answer: 12317

Example 11: Find the product: 10014 x 10008

Base: 10000

	Number		**Diff.**
	10014		+14
x	10008		+8
	10022	/	0112

RHS will have 4 digits as the base has 4 zeros.

Final answer: 100220112

5.3 When one number is above the base and one number is below the base.

Example 12: 105 x 97

As we can see, both the numbers are close to 100. Hence the appropriate base will be 100. But when we calculate the difference, one will be positive and the other will be negative.

	Number	**Difference**
	105	**+5**
X	**97**	**-3**

To find the RHS, we find the product of the differences. In this case, it will be 5 x (-3) = -15.

The cross operation will give us the LHS as 105 – 3 = 102 OR 97 + 5 = 102

But since the RHS is a negative number, we have one extra step in this case: "Multiply the LHS by the base and subtract the RHS to get the final answer"

	Number		Difference
	105		+5
X	97		-3
	102	/	-15

For getting the final answer: LHS x base – RHS

= 102 x 100 – 75

= 10200 – 75

= 10175

Final answer: 10175

Example 13: Find the product 1022 x 998

	Number	Difference
	1022	+22
X	998	-2
	1020	/ -044

Final answer: LHS x base – RHS

= 1020 x 1000 – 044

= 1020000 – 44

= 1019956

Example 14: Find the product: 980 x 1024

Base: 1000

	Number	Difference
	980	-20
X	1024	+24
	1004	/ -480

Final answer: LHS x base – RHS

= 1004 x 1000 – 480

= 1004000 – 480

= 1003520

5.4 When the numbers are closer to different bases.

Example 15: 85 x 995

In this example we can see that 85 is closer to 100 whereas 995 is closer to 1000. We will make the smaller number closer to the bigger base by multiplying it with an appropriate power of 10. In this example, if we multiply 85 by 10 then it becomes 850 which is also closer to 1000. Now we use the base method of multiplication to find 850 x 995. The answer found by this method will be divided by 10 to get the final answer.

	Number	Difference
	850	-150
X	995	-5
	845	/ 750

Answer obtained by base method is: 845750.

This will be divided by 10 to get the final answer.

Hence, 85 x 995 = 84750 ÷ 10 = 8475

Answer: 85 x 995 = 8475

Example 16: Find the product: 73 x 995

Here we see that 73 is closer to 100 and 995 is closer to 1000. So, we multiply 73 by 10 to make it 730 and then apply the base method of multiplication.

	Number		**Difference**
	730		-270
X	995		-5
	725	/	$_1$350

RHS will have 3 digits as 1000 has 3 zeros.

Answer obtained by base multiplication: 726350

Actual answer: 726350 ÷ 10 = 72635

Final answer: 73 x 995 = 72635

Summary:

Case 1: When both the numbers are below / above the same base:

1. Select the base and write the difference between the number and the base.
2. The answer will have two parts. The RHS and LHS. The RHS will have the same number of digits as the number of zeroes in the selected base.
3. RHS is obtained by multiplying the differences obtained in step 1
4. LHS is obtained by doing cross operation.

Case 2: When one number is below the base and the other is above the base:

1. Select the base and write the difference between the number and the base.
2. The answer will have two parts. The RHS and LHS. The RHS will have the same number of digits as the number of zeroes in the selected base.
3. LHS is obtained by doing cross operation.
4. RHS will be negative in this case. Therefore, we add another step for final answer: Multiply the LHS by the base and subtract the RHS to get the final answer.

Try it yourself

Q1. Find the product

(a) 82 x 93
(b) 99 x 99
(c) 94 x 91
(d) 995 x 991
(e) 994 x 996
(f) 989 x 995
(g) 975 x 992
(h) 9993 x 9995
(i) 9987 x 9988
(j) 9977 x 9989

Q2. Find the product:

(a) 105 x 108
(b) 109 x 114
(c) 102 x 111
(d) 1005 x 1004
(e) 1008 x 1014
(f) 10013 x 10008
(g) 10004 x 10011

Q3. Find the product:

(a) 87 x 103
(b) 95 x 105
(c) 93 x 109
(d) 995 x 1003
(e) 987 x 1008
(f) 998 x 1004
(g) 1011 x 988

Q4. Find the product:

(a) 87 x 991

(b) 93 x 1002

(c) 995 x 83

(d) 91 x 10004

(e) 991 x 104

(f) 995 x 108

(g) 997 x 94

(h) 1004 x 97

Chapter 6

Vertically and crosswise multiplication

In the previous chapters, we have learnt many new techniques of multiplying numbers. In chapter 3, we learnt about multiplying by 9's and in chapter 4, we learnt about multiplying by 11 and in chapter 5 we multiplied numbers near the bases. In this chapter, however, we will look at a general method for multiplication which can be applied to any set of numbers.

This method is based on the principle "*Urdhava-TiryakSutra*". It means vertically and cross wise multiplication. It can be used for any combination of numbers of any length.

6.1 Multiplication 2 x 2 digit

Example 1: 71 x 36

Traditional method:

		7	1
	X	3	6
	4	2	6
2	1	3	x
2	5	5	6

Now we will look at the Vedic method which will get us the answer in a single step.

The steps involved in the Vedic method can be understood with the help of the * diagram given below. Each * represents a digit in the multiplication and the line represents the digits to be multiplied in that step.

Step 1:

$$\begin{array}{ccc} & * & * \\ \times & * & * \end{array}$$

In this step we multiply the unit's place digit of the multiplicand with the unit place digit of the multiplier (Vertical multiplication).

We will do this step simultaneously with our example as well.

$$\begin{array}{ccc} & 7 & 1 \\ \text{X} & 3 & 6 \\ \hline & & 6 \end{array}$$

Step 2:

$$\begin{array}{ccc} & * & * \\ \times & * & * \end{array}$$

In this step we cross multiply the digits and add the products.

$$\begin{array}{ccc} & 7 & 1 \\ \text{X} & 3 & 6 \end{array}$$

So, we do (6 x 7) + (3 x 1) = 45= $_{4}5$ (4 will be carried forward to the next step.

$$\begin{array}{ccc} & 7 & 1 \\ \text{X} & 3 & 6 \\ \hline & {}_{4}5 & 6 \end{array}$$

Step 3:

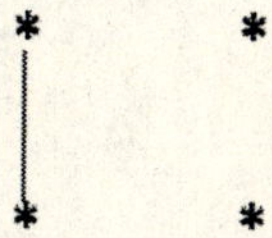

In this step we multiply the ten's digit of the multiplicand with the ten's digit of the multiplier (Vertical multiplication)

		7	1
	X	3	6
2	5	5	6

In this step we do (7x 3) + 4 (carry forward) = 25

The three steps can be summarized below:

Step 1	Step 2	Step 3

Vertically multiply the unit's place column	Cross multiply the two columns and add the products	Vertically multiply the ten's place column

NOTE: If the answer at any step is a two-digit answer, we write down only one digit and carry over the other digits to the next step.

Let us look at few other 2 x 2 examples.

Example 2: Find the product: 34 x 56

	3	4
X	5	6

Step 1: 6 x 4 = 24 = $_{2}4$

Step 2: (6 x 3) + (5 x 4) + 2 = 18 + 20 + 2 = $_{4}0$

Step 3: (5 x 3) + 4 = 15 + 4 = 19

Answer: 1904

Example 3: 84 x 47

	8	4
X	4	7
3 9	$_{7}4$	$_{2}8$

Step 1: 7x 4 = 28 = $_{2}8$

Step 2: 7 x 8 + 4 x 4 + 2 = 74 = $_{7}4$

Step 3: 8 x 4 + 7 = 39

Answer: 3948

6.2 Multiplication 3 x 3 digits

Let us try to understand the 3 x3 digit multiplication through a * diagram.

Steps for 3 x 3 multiplication

Step 1:

Multiply vertically in the unit's place column

Step 2:

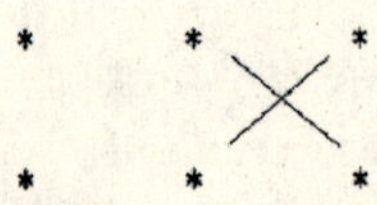

Multiply cross wise the ten's place and the unit's place columns and add the products.

Step 3:

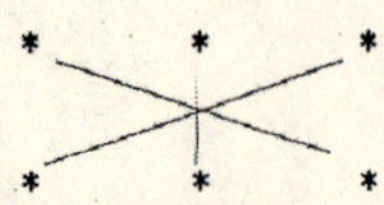

Multiply cross wise the hundred's place and the unit's place AND vertically the ten's place and add all the three products.

Step 4:

Multiply cross wise the ten's and the hundred's place columns and add the products.

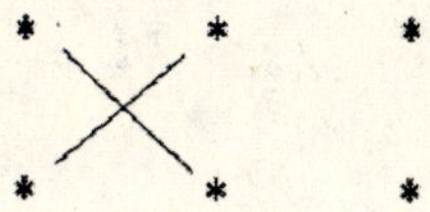

Step 5:

Multiply vertically the hundred's place column

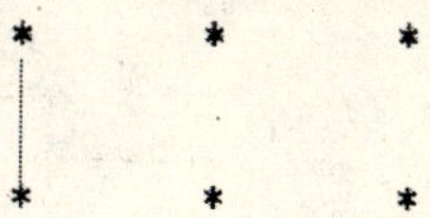

The five steps in the 3 x 3 multiplication are summarized below

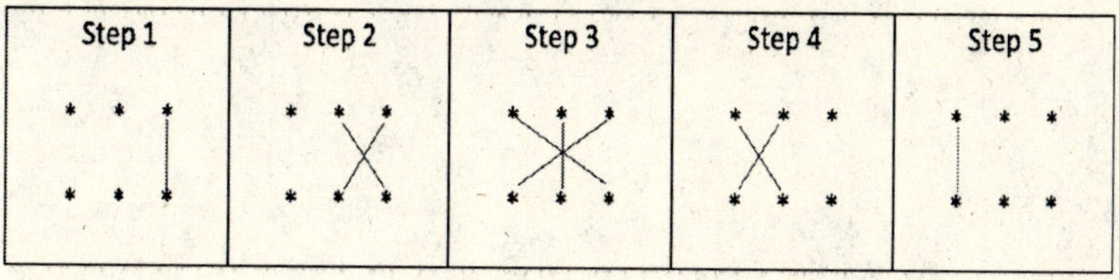

Example 4: 413 x 234

Step 1:

	4	1	3	Vertically – unit's place column
x	2	3	4	$3 \times 4 = 12 = {}_{1}2$
			${}_{1}2$	

Step 2:

		4	1	3
	x	2	3	4
			$_{1}4$	2

Cross wise- ten's & unit's place columns

$(4 \times 1) + (3 \times 3) + 1 = 14 = {}_{1}4$

Step 3:

		4	1	3
	x	2	3	4
		$_{2}6$	4	2

Vertically and cross wise

$(4 \times 4) + (2 \times 3) + (3 \times 1) + 1 = 26 = {}_{2}6$

Step 4:

		4	1	3
	x	2	3	4
	$_{1}6$	6	4	2

Cross wise – ten's & hundred's columns

$(3 \times 4) + (2 \times 1) + 2 = 16 = {}_{1}6$

Step 5:

		4	1	3
	x	2	3	4
9	6	6	4	2

Vertically the hundred's column

$(2 \times 4) + 1 = 9$

Answer: 96642

Example 5: Find the product: 347 x 812

Step 1:

	3	4	7
x	8	1	2
			$_{1}4$

$(2 \times 7) = 14 = {}_{1}4$

Step 2:

		3	4	7
	x	8	1	2
			$_{1}6$	4

$(2 \times 4) + (1 \times 7) + 1 = 16 = {}_{1}6$

Step 3:

		3	4	7
	x	8	1	2
		$_{6}7$	6	4

$(2 \times 3) + (8 \times 7) + (1 \times 4) + 1 = 67 = {}_{6}7$

Step 4:

		3	4	7
	x	8	1	2
	$_{4}1$	7	6	4

$(1 \times 3) + (8 \times 4) + 6 = 41 = {}_{4}1$

Step 5:

		3	4	7
	x	8	1	2
28	1	7	6	4

$(8 \times 3) + 4 = 28$

Answer: 281764

Example 6: Find the product: 753 x 924

		7	5	3
	x	9	2	4
69	$_{6}5$	$_{6}7$	$_{2}7$	$_{1}2$

Answer: 695772

6.3 Multiplication 4 x 4 digits

Let us take a look at the * diagram for a 4 x 4-digit multiplication. We will also solve an example simultaneously to understand the steps better.

Example 7: 3254 x 1467

Step 1: Vertically unit's place column

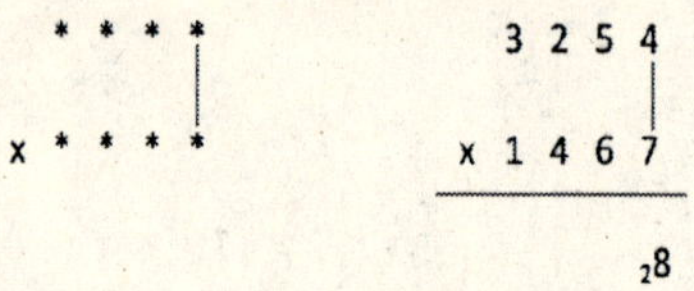

Step 2: Cross wise unit's & ten's place column

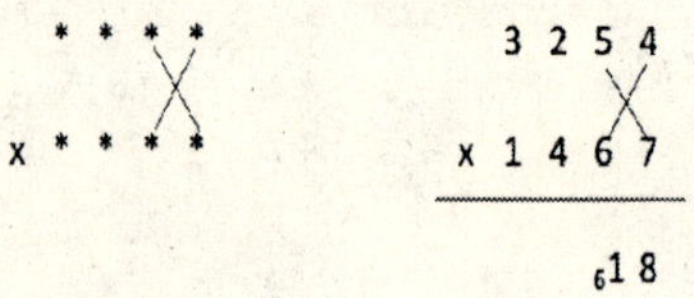

Step 3: Cross wise and Vertical multiplication

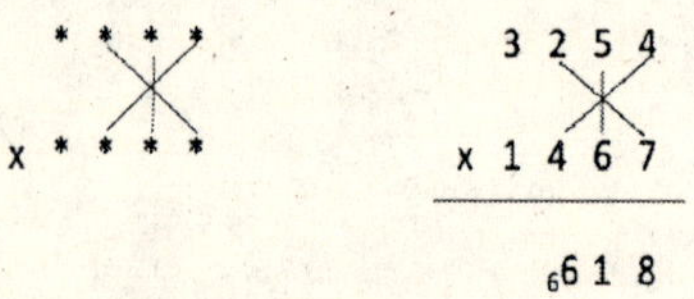

Step 4: Cross wise and cross wise multiplication

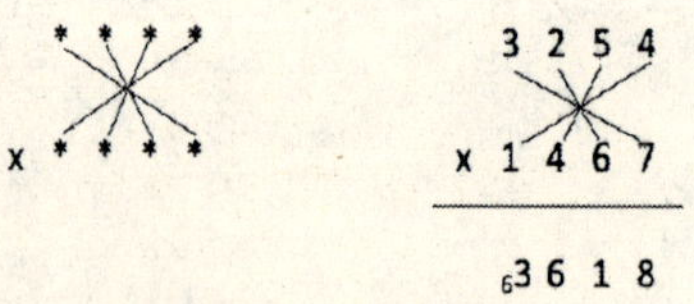

Step 5: Cross wise and vertical multiplication

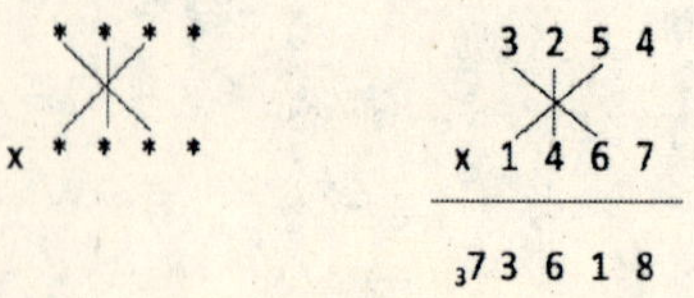

Step 6: Cross wise multiplication:

```
  * * * *          3 2 5 4
   X                 X
x * * * *        x 1 4 6 7
                 ---------
               ₁7 7 3 6 1 8
```

Step 7: Vertical multiplication

```
  * * * *          3 2 5 4
  |                |
x * * * *        x 1 4 6 7
                 ---------
               4 7 7 3 6 1 8
```

Answer: 3254 x 1467 = 4773618

Let us look at one more example.

Example 8: Find the product: 7101 x 3072

Step 1:

```
    7 1 0 1
          |
  x 3 0 7 2
  ---------
          2
```

Step 2:

```
    7 1 0 1
         X
  x 3 0 7 2
  ---------
        7 2
```

Step 3:

```
    7 1 0 1
       \|/
       /|\
  x 3 0 7 2
  ---------
      2 7 2
```

Step 4:

$$\begin{array}{r} 7\ 1\ 0\ 1 \\ \times\ 3\ 0\ 7\ 2 \\ \hline {}_{2}4\ 2\ 7\ 2 \end{array}$$

Step 5:

$$\begin{array}{r} 7\ 1\ 0\ 1 \\ \times\ 3\ 0\ 7\ 2 \\ \hline {}_{5}1\ 4\ 2\ 7\ 2 \end{array}$$

Step 6:

$$\begin{array}{r} 7\ 1\ 0\ 1 \\ \times\ 3\ 0\ 7\ 2 \\ \hline 8\ 1\ 4\ 2\ 7\ 2 \end{array}$$

Step 7:

$$\begin{array}{r} 7\ 1\ 0\ 1 \\ \times\ 3\ 0\ 7\ 2 \\ \hline 218\ 1\ 4\ 2\ 7\ 2 \end{array}$$

Answer: 21814272

6.4 Characteristics of Vedic Multiplication

Let us look at the * patterns for 2 x 2, 3 x 3 and 4 x 4 multiplication again and understand the characteristics of Vedic Multiplication.

	Step 1	Step 2	Step 3	Step 4	Step 5	Step 6	Step 7
2 x 2	* * * *	* * * *	* * * *				
3 x 3	* * * * * *	* * * * * *	* * * * * *	* * * * * *	* * * * * *		

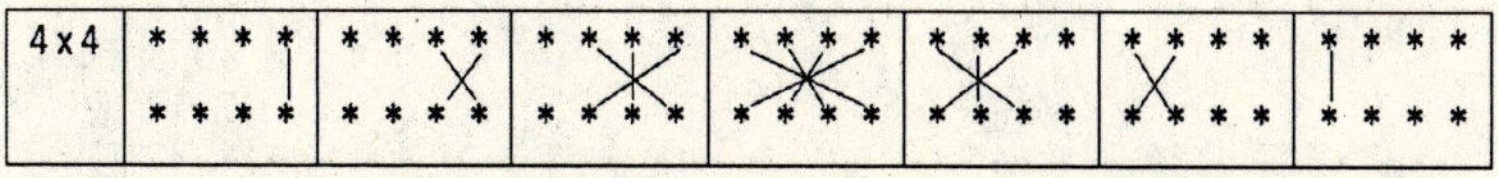

Looking at the table above, we observe the following characteristics:

1. If the number of digits is same in the multiplicand and the multiplier then the number of steps needed to complete the multiplication can be given by the equation
   Number of steps = 2 x number of digits – 1
   So, for 2-digit multiplication, we need 2x 2 – 1 = 3 steps
   For a 3-digit multiplication we need 2 x3 – 1 = 5 steps.
   Note: If there are unequal number of digits in the multiplicand and the multiplier, they should first be made equal by inserting leading zeros. For example, the multiplication of 634 x 23 can be written as 634 x 023 and then the steps of 3 x 3 multiplication applied.
2. The number of steps involved in Vedic multiplication is always odd. Among these odd steps, the first step and the last step are mirror images of each other. Similarly, the second step and the second last step are mirror images of each other and so on.
3. In every case there is a central step which involves all the digits and it does not have a mirror image.
4. We always start with the digits in the unit's place and keep adding one column in each next step till all the columns are involved in the multiplication, which is the central step. Then, we start leaving out one column from the right in each step till we reach the left most column.

6.5 Multiplication 5 x 5 digits

Based on the characteristics listed in section 6.4, we can now make the * diagram for a 5 x 5 multiplication.

As seen from characteristic 1, the number of steps for a 5 x5 digit multiplication will be 2x5 – 1 = 9 steps

Step 1:

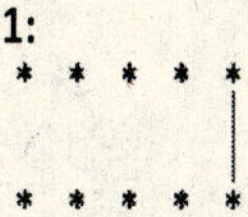

Step 2:

Step 3:

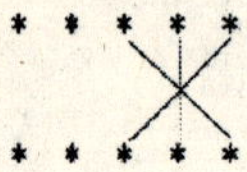

Step 4:

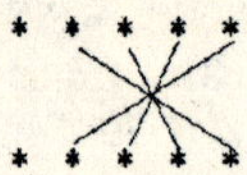

Step 5:

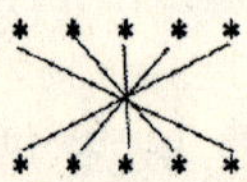

Step 6:

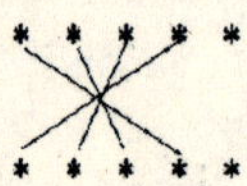

Step 7:

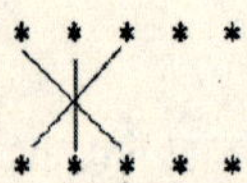

Step 8:

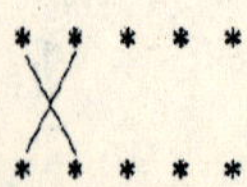

Step 9:

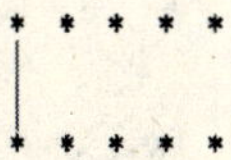

Example 9: Find the product: 23145 x 52301

$$\begin{array}{r} 2\;3\;1\;4\;5 \\ \times\; 5\;2\;3\;0\;1 \\ \hline \end{array}$$

Step 1:

$$\begin{array}{r} 2\;3\;1\;4\;5 \\ \times\; 5\;2\;3\;0\;1 \\ \hline 5 \end{array}$$

Step 2:

$$\begin{array}{r} 2\;3\;1\;4\;5 \\ \times\; 5\;2\;3\;0\;1 \\ \hline 4\;5 \end{array}$$

Step 3:

$$\begin{array}{r} 2\;3\;1\;4\;5 \\ \times\; 5\;2\;3\;0\;1 \\ \hline {}_{1}6\;4\;5 \end{array}$$

Step 4:

$$\begin{array}{r} 2\;3\;1\;4\;5 \\ \times\; 5\;2\;3\;0\;1 \\ \hline {}_{2}6\;6\;4\;5 \end{array}$$

Step 5:

```
      2 3 1 4 5
    x 5 2 3 0 1
  -------------
     ₄0 6 6 4 5
```

Step 6:

```
      2 3 1 4 5
    x 5 2 3 0 1
  -------------
   ₃5 0 6 6 4 5
```

Step 7:

```
        2 3 1 4 5
      x 5 2 3 0 1
    -------------
   ₂0 5 0 6 6 4 5
```

Step 8:

```
          2 3 1 4 5
        x 5 2 3 0 1
      -------------
   ₂1 0 5 0 6 6 4 5
```

Step 9:

```
             2 3 1 4 5
           x 5 2 3 0 1
         -------------
   121 0 5 0 6 6 4 5
```

Answer: 1210506645

Example 10: Find the product: 56017 x 90715

$$\begin{array}{r} 5\ 6\ 0\ 1\ 7 \\ \times\ 9\ 0\ 7\ 1\ 5 \\ \hline \end{array}$$

$50\ {}_58\ {}_41\ {}_65\ {}_98\ {}_42\ {}_51\ {}_15\ {}_35$

Answer: 5081582155

Summary:

1. The multiplicand and multiplier should have same number of digits. If the number of digits is not same, add leading zeros to make the digits same.
2. The number of steps required is always given by the equation:
   $number\ of\ steps = 2n - 1$, where n is the number of digits in the multiplicand / multiplier.
3. Begin at the right most column (unit's place column) and moving left, keep on including one column at a time till you reach the step where all the columns of the numbers are involved. Then keep excluding one column from the right till you reach the left most column.
4. The number of steps should always be odd.

	Step 1	Step 2	Step 3	Step 4	Step 5	Step 6	Step 7
2 x 2							
3 x 3							
4 x 4	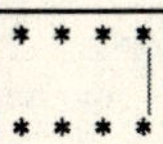	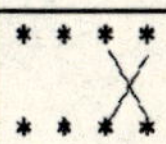	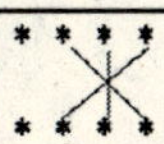	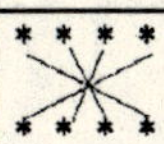	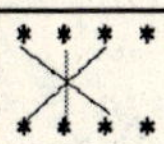	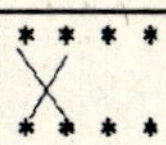	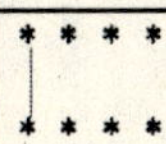

Try it yourself

Q1. Find the product:

(a) 35 x 78
(b) 97 x 34
(c) 56 x 42
(d) 79 x 82
(e) 85 x 29

Q2. Find the product:

(a) 345 x 679
(b) 508 x 347
(c) 796 x 278
(d) 654 x 198
(e) 432 x 549

Q3. Find the product:

(a) 3258 x 5026
(b) 7079 x 3451
(c) 4127 x 6807
(d) 6581 x 4532
(e) 6754 x 3021

Chapter 7

Division – Base method Part 1

7.1 Division: Key terms

Division is splitting into equal parts or groups. For example, if we have 12 apples and we want to split them equally between 4 children, each child will get 3 apples. The key vocabulary terms for division are:

Dividend – The quantity or the number to be divided. 12 is the dividend in the above example.

Divisor – the number of parts or groups into which the dividend is to be split. 4 is the divisor in the above example.

Quotient – The result of division or split. 3 is the quotient in the given example.

The symbol for division is (÷) or (/).

Division is the reverse of multiplication. If we know a multiplication fact, we can write the corresponding division fact. For example,

$4 \times 3 = 12$ is a multiplication fact. This one fact of multiplication gives us two division facts

$12 \div 4 = 3$ and $12 \div 3 = 4$

But sometimes, the split is not fair. For example, if we have 9 sweets to be split among 4 children, then each child will get 2 sweets each, but 1 sweet will be left over. This left over is called '***remainder***'.

7.2 Base Method of division

In chapter 5, we learnt about the base method of multiplication. Base method is used when the numbers are near the base. For division also, base method is used when the divisor is close to a base. We will be using the following schematic diagram to explain division by base method:

Base

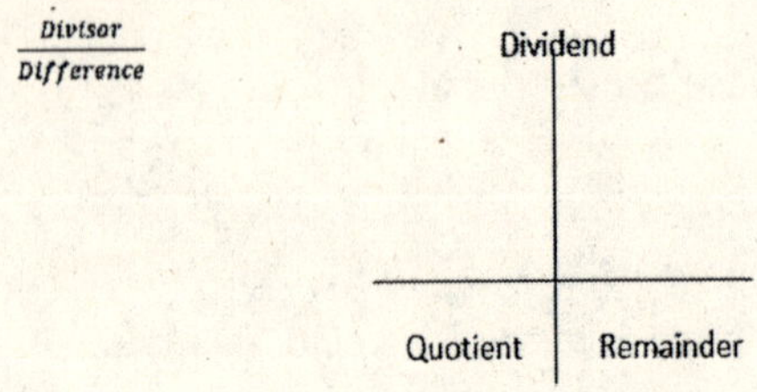

Figure 7.1

The dividend will be split into two parts: the left-hand side (LHS) and the right-hand side (RHS). The RHS will have the same number of digits as the zeros in the selected base.

We will be categorizing the division problems based on the divisor.

1. When the divisor is below the base
2. When the divisor is above the base (Chapter 8)

7.3. When the divisor is below the base and the base is 10.

Example 1: 233 ÷ 9

Step 1: Selecting the base and finding the difference.

In this question, the divisor is 9, which is close to 10. Hence the base selected will be 10. The difference between the base and the number is 1.

Step 2: Splitting the dividend: The base has 1 zero. Based on this, the dividend will be split up in a such a way that the RHS of the dividend has 1 digit.

Using the diagram in figure 7.1, we get,

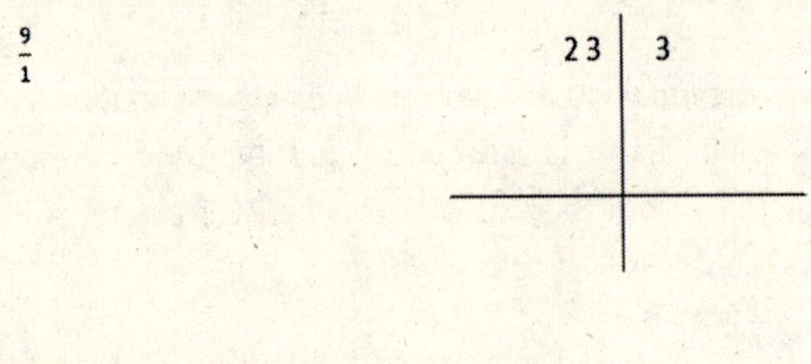

Step 3: Bring down the first digit of the dividend, as shown. This becomes our first quotient digit.

Step 4: Multiply 2 with the difference 1 and write the answer below the next digit of the dividend, which is 3 in this case.

2 x 1 = 2 and it is written below 3 and added to it.

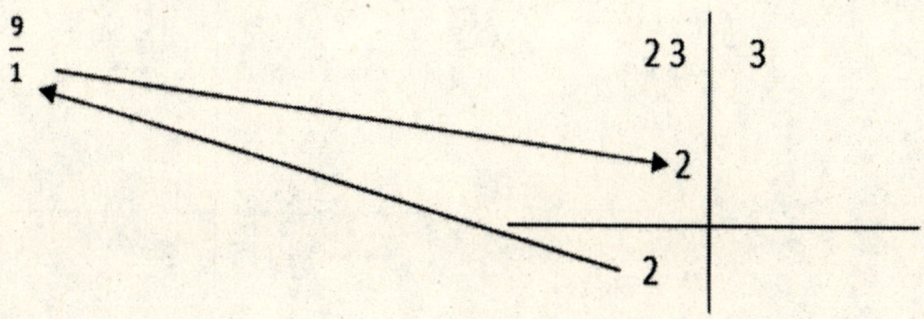

Step 5:

Add both the digits 3 and 2 in the second column to get 5 which is our second quotient digit.

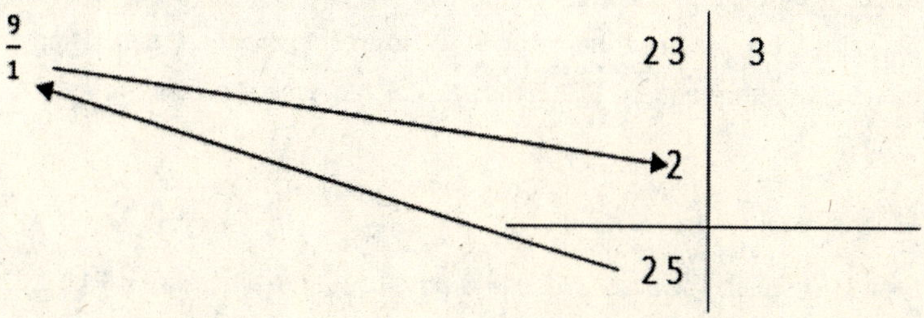

Step 6: We now multiply this quotient digit, 5, with the difference and write the product in under the next digit of the dividend, that is below 3, which is on the RHS of the dividend.

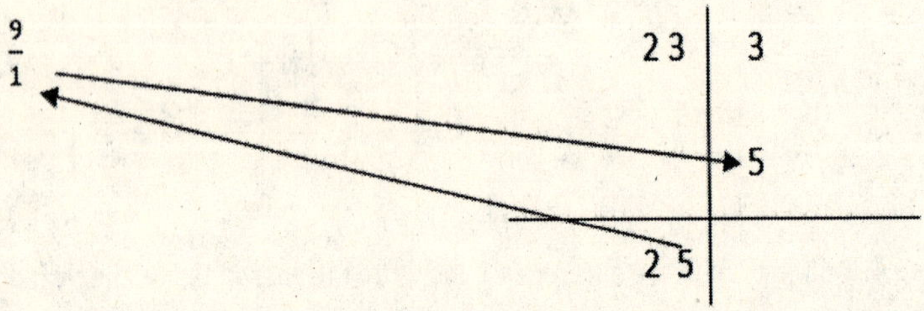

Step 7: Finally add the numbers in the remainder column to get 8.

Divisor / difference	Quotient side	Remainder
$\frac{9}{1}$	23	3
		5
	25	8

Final answer: 233 ÷ 9 gives 25 as quotient and 8 as remainder.

Important point when dividing by 9

When we divide by 9, the difference is 1. So, when we find the product of the quotient digit and the difference, the answer is always going to be the quotient digit. So, when we are dividing by 9, we can eliminate the product step and directly write the quotient digit in the next column. We can see this from the next example.

Example 2: 161 ÷ 9

Step 1: We select the base as 10 as 9 is close to 10. The difference will be 1.

Step 2: The dividend will be split into two parts such that the RHS has 1 digit as the base has 1 zero.

10

$\frac{9}{1}$

	1 6	1
	1	7
	1 7	8
	Quotient	Remainder

Answer: 161 ÷ 9 gives 17 as quotient and 8 as remainder.

Example 3: 1241 ÷ 9

Step 1: Base is 10 and the difference is 1.

$\frac{9}{1}$

	1 2 4	1
	Quotient	Remainder

Step 2: Bring down 1 as it is in the quotient column.

10

$\frac{9}{1}$

	1 2 4	1
	1	

Step 3: add this quotient digit to the next dividend digit to get (2 + 1 = 3) as the next quotient digit.

Divisor	Dividend	Remainder
10		
9 / 1	1 2 4	1
	1	
	1 3	

Step 4: Now add this quotient digit 3 to the next dividend column to get (4 + 3 = 7) as the next quotient digit.

Divisor	Dividend	Remainder
10		
9 / 1	1 2 4	1
	1 3	
	1 3 7	

Step 5: Now add this third quotient digit to the next dividend digit which is 1 in the remainder column. This will give us (1 + 7= 8) This 8 is our final remainder.

Divisor	Dividend	Remainder
10		
9 / 1	1 2 4	1
	1 3	7
	1 3 7	8

Answer: 1241 ÷ 9 gives 137 as quotient and 8 as remainder.

Example 4: 502 ÷ 9

10

9
—
1

5 0	2
5	5
5 5	7

Answer: 502 ÷ 9 will give 55 as quotient and 7 as remainder.

NOTE: In division, the remainder is always less than the divisor. It is possible that the addition in the remainder column yields a number which is greater than the divisor. In this case we will further divide the remainder by the divisor and add the new quotient to the quotient obtained by the base method. This will be explained with the help of the following example.

Example 5: 536 ÷ 9

Step 1: Base is 10 and the difference is 1

Step 2: Split the dividend into two parts such that the RHS contains only one digit.

10

9
—
1

5 3	6
Quotient	Remainder

Step 2: Bring down the first digit in the quotient column as it is.

10

9
—
1

5 3	6
5	

Step 3: Add this quotient digit to the next digit in the dividend column to get (3 + 5= 8). This is our next quotient digit.

10		
$\frac{9}{1}$	5 3	6
	5	
	5 8	

Step 4: Now add this 8 to the next digit in the dividend which is 6 in the remainder column. We will get (8+ 6= 14).

10		
$\frac{9}{1}$	5 3	6
	5	
	5 8	14

Now 14 is greater than 9. So, we further divide 14 by 9. It gives 1 as the quotient and 5 as the remainder. This 1 is added to the quotient obtained in the previous step and we get our final quotient as 58 + 1 = 59 and our remainder as 5.

Final Answer: 502 ÷ 9 gives us 59 as quotient and 5 as remainder.

Example 6: 249 ÷9

10		
$\frac{9}{1}$	2 4	9
	2	6
	2 6	15

As we can see, the quotient obtained by base method is 26 and the remainder is 15 which is greater than the divisor 9. Hence, we further divide 15 by 9 and get quotient as 1 and the remainder as 6. Adding this quotient to the quotient obtained by base method, we get our final quotient as 27 and the remainder as 6.

Final answer: 249 ÷ 9 gives 27 as quotient and remainder as 6.

7.4 Divisors other than 9

Example 7: 231 ÷8

Step 1: Selecting the base and finding the difference: In this case the base is 10 and the difference between the base and the divisor is 2.

Step 2: Split the dividend into two parts such that the RHS contains one digit.

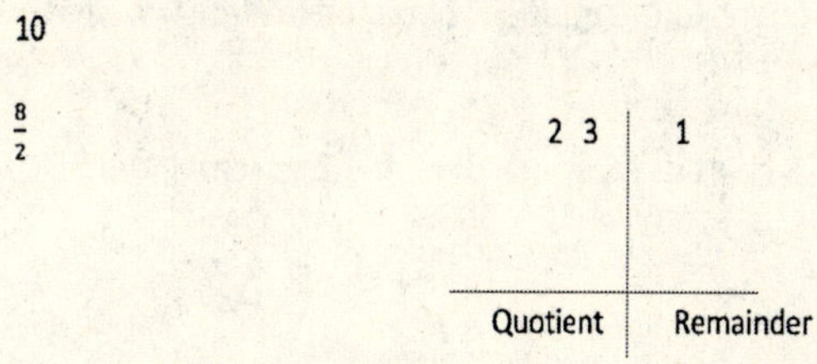

Step 3: Bring down the first digit of the dividend as it is. This will give us our first digit of quotient. 10

Step 4: Multiply this digit of the quotient to the difference and add the product to the second digit of the dividend. This will give us (2 x 2) + 3 = 7 as the second digit of the dividend. 10

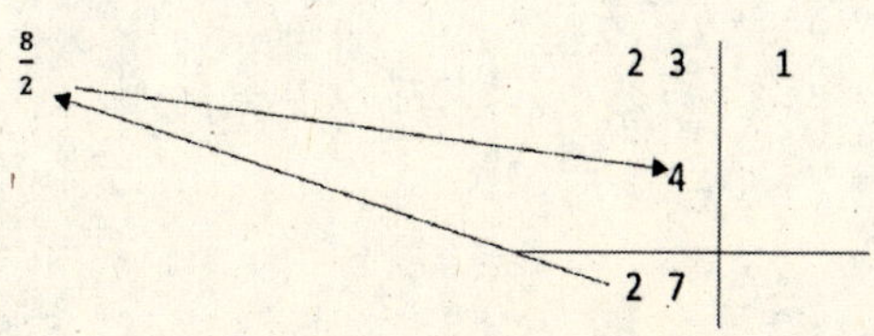

Step 5: Multiply 7 with the difference and add the product to the last digit of the dividend, 1, which is in the remainder column. So we get (1 + 14 = 15).

10		
$\frac{8}{2}$	2 3	1
	4	14
	2 7	15

Step 6: Now 15 is greater than the divisor. So, we further divide the remainder, 15, by 8 which gives us quotient as 1 and the remainder as 7.

Adding this quotient to the quotient obtained in step 4, we get 27 + 1 = 28 and remainder as 7.

Answer: 231 ÷ 8 gives us 28 as quotient and 7 as remainder.

Example 8: 121 ÷ 7

Step 1: The base is 10 and the difference is 3.

Step 2: Split the dividend into two parts such that the RHS has one digit.

10		
$\frac{7}{3}$	1 2	1

Step 3: Bring down the first digit of the dividend as it is. This gives us the first digit of the quotient.

10		
$\frac{7}{3}$	1 2	1
	↓	
	1	

Step 4: Multiply this digit, 1, by the difference and add the product to the second digit of the dividend. We get (1 x 3) + 2 = 5.

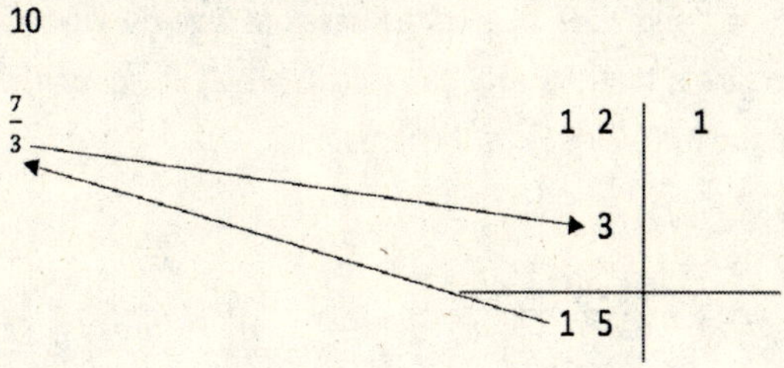

Step 5: Multiply 5 with the difference and add the product to the third digit of the dividend. We get (5 x 3) + 1 = 16

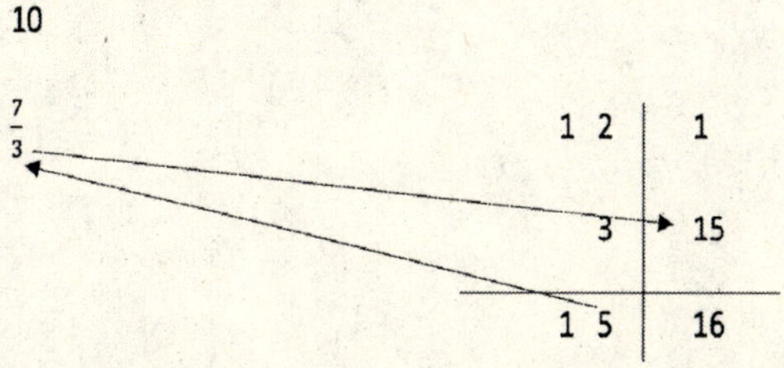

Step 6: We see that 16 is greater than the divisor. So we divide it further by 7 to get quotient as 2 and the remainder as 2. Adding this quotient to the quotient obtained in step 5, we get our final quotient as 17 and remainder as 2.

Answer: 121 ÷7 gives 17 as quotient and 2 as remainder.

7.5 Divisors closer to base 100.

For divisors like 99, 98, 97 the selected base will be 100 which has two zeros. In this case we will split the divided into two parts such that the RHS has 2 digits. Another important point to remember is that the difference between the base and the number will be written in two digits. For example, if the divisor is 99, the difference will be 01. And when we find the product at each step between the quotient and the difference, that product will also be written as a 2-digit answer.

Let us understand this better with the help of an example:

Example 9: 1702 ÷ 99

Step 1: Selecting the base and finding the difference. The base selected will be 100 and the difference will be 01.

Step 2: Splitting the dividend: Since the base has 2 zeros, we split the dividend into two parts such that the RHS has 2 digits. The LHS column will give us the quotient and the RHS will give us the remainder.

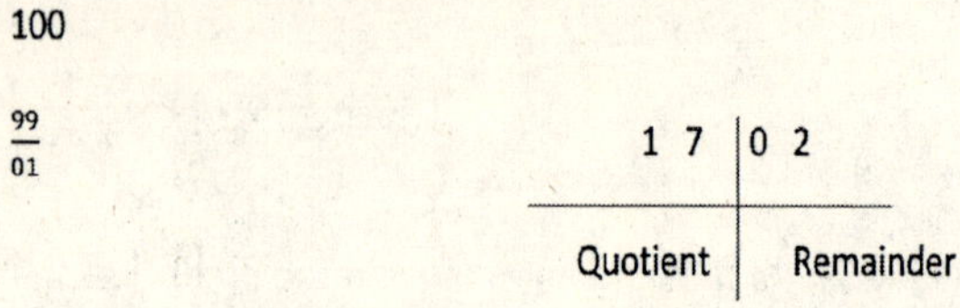

Step 3: Bring down the first digit of the dividend as it is. This gives us the first digit of the quotient

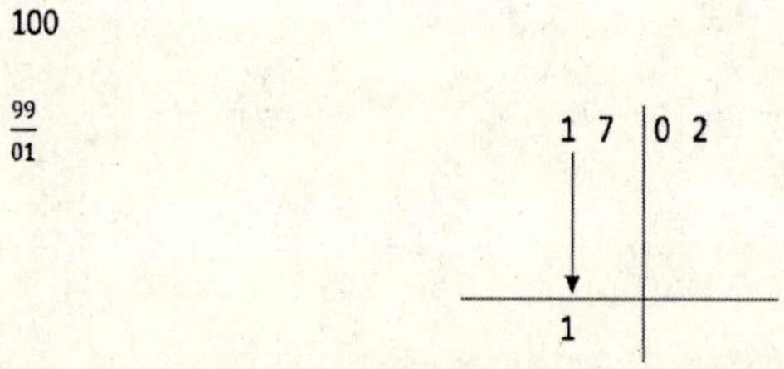

Step 4: Multiply this digit with the difference and add the product to the next digits in the dividend. Since this product will be a 2-digit number, it will be written under 2 digits of the dividend as shown below. (1 x 01) = 01. Write this below the next two digits, 7 and 0, of the dividend. Add the digits in the second column of the dividend to get 7 as the second digit of the quotient.

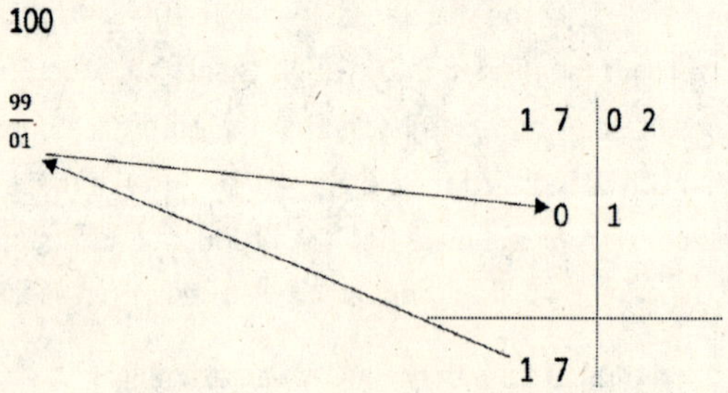

Step 5: Multiply the second digit of the quotient, 7, to the difference and write the product under the next two digits of the dividend, 0 and 2.

Product is 7 x 01 = 07. This will be written under 0 and 2.

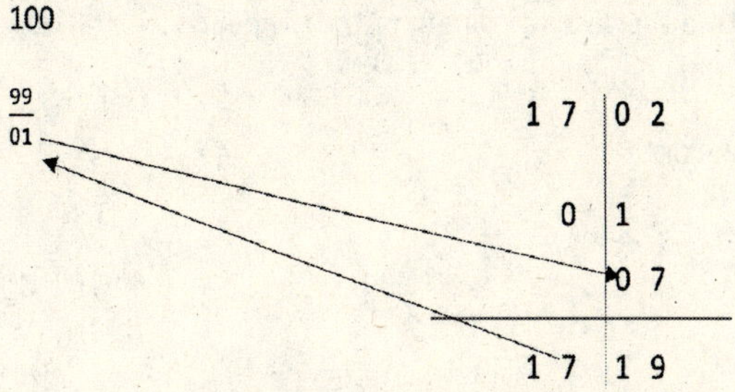

Step 6: Adding the digits in the remainder column we get 19 as the remainder.

Answer: 1702 ÷ 99 gives 17 as the quotient and 19 as the remainder.

Example 10: 2405 ÷ 98

Step 1: Base is 100. Difference is 02.

Step 2: Split the dividend into two parts such that the RHS has 2 digits.

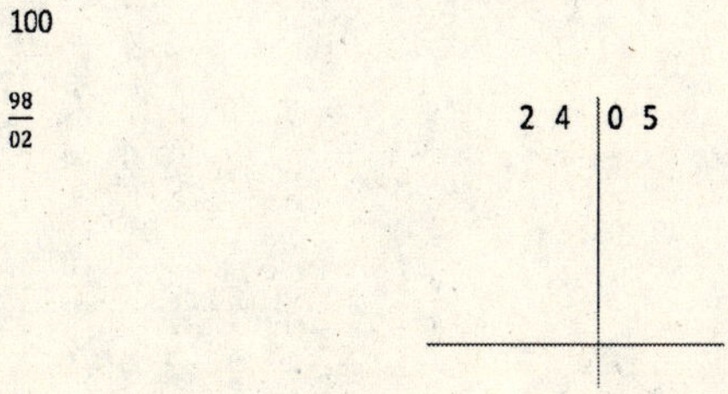

Step 3: Take down the first digit of the dividend as it is. This becomes the first digit of the quotient.

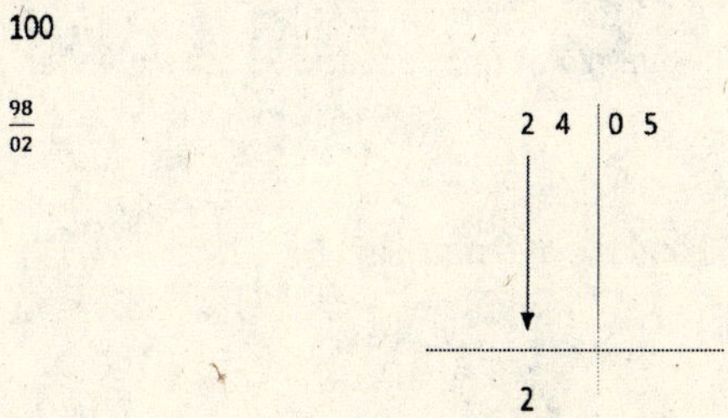

Step 4: Multiply this digit, 2, with the difference and write the product under the next two digits of the dividend, 4 and 0.

2 x 02 = 04

Add the digits in the second column of the dividend to get 4 + 0 = 4. This is the second digit of the quotient.

100

$\frac{98}{02}$

2 4	0 5
0	4
2 4	

Step 5: Multiply 4 with the difference to get 08. Write this product under the next two digits of the dividend, 0 and 5.

Add the columns under the remainder section to get.

100

$\frac{98}{02}$

2 4	0 5
0	4
	0 8
2 4	5 3

Note that when we add the digits in the remainder column, it is conventional addition with carry over.

Answer: 2405 ÷ 98 gives 24 as quotient and 53 as remainder.

Example 11: 5037 ÷ 89

Step 1: Base is 100 and the difference is 11

Step 2: Splitting the dividend

100

$\frac{89}{11}$

5 6	3 7
Quotient	Remainder

Step 3:

100

$\frac{89}{11}$

5 6 ↓	3 7
5	

Step 4:

100

$\frac{89}{11}$

5 6	3 7
5	5
5 $_1$1	

Step 5:

100

$\frac{89}{11}$

5 6	3 7
5	5
	$_1$2 1
6 1	208

Note: In this step when we multiplied 11 with the difference 11, we get 121. It has three digits, so it will be written under two digits of the remainder column of the dividend with 1 in 121 as carry over. Adding the remainder column we get, 208

Divisor	Quotient column	Remainder column
100		
$\frac{89}{11}$	5 6	3 7
	5	5
		$_{1}2$ 1
	6 1	208

Step 6: 208 is greater than the divisor. So, it will be further divided by 89 to get quotient as 2 and the remainder as 30. Adding this quotient to the quotient obtained in the previous step we get 63 as quotient and 30 as remainder.

Divisor	Quotient column	Remainder column
100		
$\frac{89}{11}$	5 6	3 7
	5	5
		$_{1}2$ 1
	6 3	30

Some more examples:

Example 12: 1345 ÷ 97

Divisor	Quotient column	Remainder column
100		
$\frac{97}{03}$	1 3	4 5
	0	3
		0 9
	1 3	8 4

Answer: 1345 ÷ 97 gives 13 as quotient and 84 as remainder.

Example 13: 6312 ÷ 95

100		
95	6 3	1 2
05	3	0
		3 0
	6 6	4 2

Answer: 6312 ÷ 95 gives us 66 as quotient and 42 as remainder.

Example 14: 4621 ÷ 91

100		
91	4 6	2 1
09	3	6
		8 1
	4 9	162

In this case, since the remainder is greater than the divisor, it will be further divided by 91 to give 1 as quotient and 71 as remainder. So, our final answer will be:

4621 ÷91 gives 50 as quotient and 71 as remainder.

7.6 When the divisors are close to 1000.

Since the selected base will have three zeros, the dividend will be split into two parts such that the RHS has 3 digits.

Example 14: 12315 ÷ 999

Step 1: Selected base is 1000 and the difference of the divisor from the base is 001.

Note that the difference will be written as 3 digits.

Step 2: The dividend will be split into two parts such that the RHS has 3 digits.

1000

999
001

1 2	3 1 5
Quotient	Remainder

Step 3: Bring down the first digit of the dividend as it is. That becomes the first digit of the quotient.

1000

999
001

1 2	3 1 5
1	

Step 4: Multiply this digit, 1, with the difference and write the product under the next three digits of the dividend. Adding the column in the LHS gives us the second digit of the quotient, which is 2.

1000

999
001

1 2	3 1 5
0	0 1
1 2	

Step 5: Multiply the second digit of the quotient, 2, with the difference and write the product under the next three digits of the dividend.

1000

$\frac{999}{001}$

1 2	3 1 5
0	0 1
	0 0 2
1 2	

Step 6: Adding the digits on the RHS we get:

1000

$\frac{999}{001}$

1 2	3 1 5
0	0 1
	0 0 2
1 2	3 2 7

Answer: 12315 ÷ 999 gives 12 as quotient and 327 as remainder.

Example 15: 30467 ÷ 995

1000

$\frac{995}{005}$

3 0	4 6 7
0	1 5
	0 0 0
3 0	6 1 7

Answer: 30467 ÷ 995 gives 30 as quotient and 617 as remainder.

IMPORTANT NOTE:

For any given question, we can self-check our answer by using the formula:

$$dividend = quotient \times divisor + remainder$$

For example, in our last question, we have

Dividend = 30467; Divisor = 995; Quotient = 30; Remainder = 617

Using the formula, we get,

995 x 30 + 617

= 29850 + 617

= 30467 = dividend

Hence, our calculations are correct.

Summary:

- Select the base and calculate the difference between the Base and the divisor
- Split the dividend into two parts such that the RHS has the same number of digits as the number of zeros in the selected base. The LHS of the split dividend will give us the quotient and the RHS will give us the remainder.
- Bring down the first dividend digit. This becomes the first quotient digit.
- Multiply this quotient digit with the difference calculated in the first step and write it below the next digits of the dividend. Write one digit in one column. Also the number of digits to be written at each step should be equal to the number of zeros in the selected bae.
- Add the digits in the next column to get the next digit of the quotient.
- This quotient digit is again multiplied with the difference and the product written under the next digits of the dividend.
- This process continues till we have written a digit under the last column in the RHS of the dividend.
- Adding the digits in the remainder column will give us the remainder.

- Adding the digits in the remainder column will give us the remainder.
- If the remainder is greater than the divisor, then it is further divided by the divisor and the new quotient is added to the quotient obtained earlier. Remainder obtained in this step will be the final remainder.

Try it yourself

Q1. Find the quotient and the remainder:

(a) 732 ÷ 9

(b) 631 ÷ 8

(c) 403 ÷ 9

Q2. Find the quotient and the remainder:

(a) 2134 ÷ 99

(b) 3476 ÷ 97

(c) 6087 ÷ 91

(d) 12341 ÷ 96

(e) 40768 ÷ 89

Q3. Find the quotient and the remainder

(a) 34512 ÷ 999

(b) 46081 ÷ 997

(c) 20871 ÷ 991

Chapter 8

Division – Base method Part 2

In the last chapter we learnt about dividing by numbers which are close to the base but below it. In this chapter we shall learn division when the divisor is close to the base but above it.

When we divide any number by a divisor which is just above a base, we use the formula "Transpose and Apply". The working rule is same as that for a divisor below the base but the difference between the base and the number in this case will be negative and it will be written with a negative sign.

The product that is obtained by quotient x difference will be written using the appropriate sign as the difference is negative.

We will be taking three cases:

(a) When the base is 10

(b) when the base is 100

(c) when the base is 1000

8.1 When the base is 10.

Example 1: 1324 ÷ 12

We will be setting up the problem just like we did in the previous chapter.

Base

$\frac{Divisor}{Difference}$ Dividend

Quotient | Remainder

Step 1: Selecting the base and finding the difference: The divisor is 12 which is close to 10. Hence the selected base will be 10 and the difference between the base and the divisor will be (-2).

Step 2: Splitting the dividend: Since the selected base has only one zero, we will split the dividend into two parts such that the RHS has only one digit.

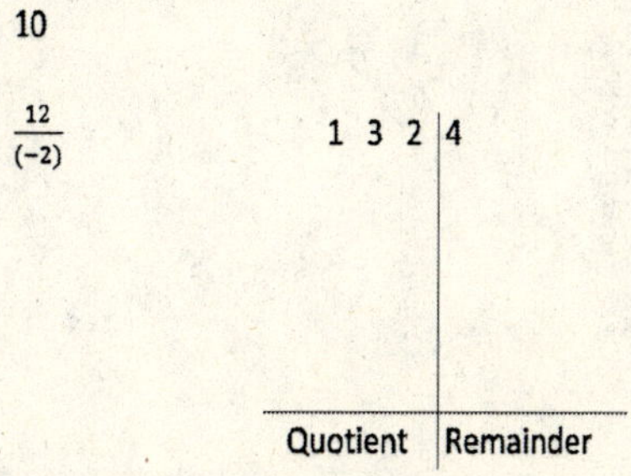

Step 3: Bring down the first digit of the dividend as it is. This becomes the first digit of the quotient.

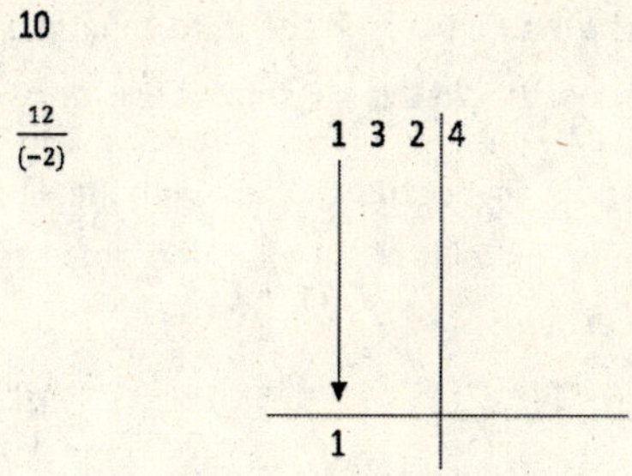

Step 4: Multiply this digit with the difference and write the product under the next digit of the dividend.

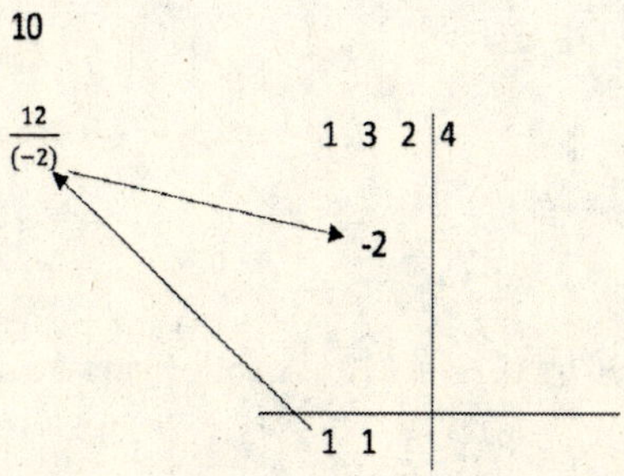

In this case, it will be 1 x (-2) = -2. This -2 will be written under the digit 3 of the dividend. Now we add the numbers in the second column of the dividend to get the second digit of the quotient. In this question, it will be 3 + (-2) = 1.

Step 5: Multiply the second digit of the quotient with the difference. Write the product under the next digit, 2, of the dividend.

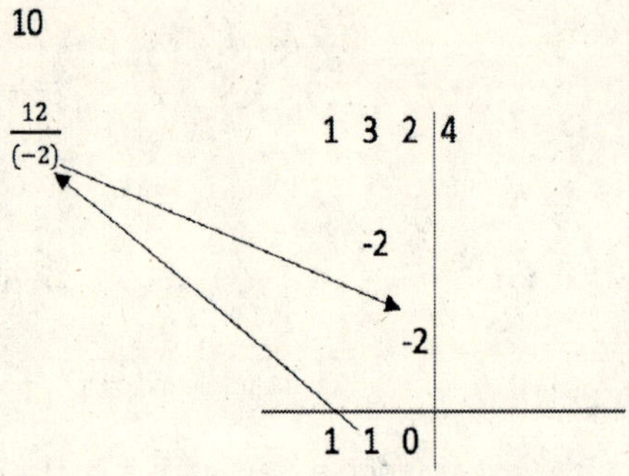

In this case, it will be 1 x (-2) = -2. This -2 will be written under the digit 2 of the dividend. Now we add the numbers in the third column of the dividend to get the third digit of the quotient. In this question, it will be 2 + (-2) = 0.

Step 6: Multiply this third digit of the quotient with the difference and write the product under the next digit, 4, of the dividend. Add this column to get the remainder for the question.

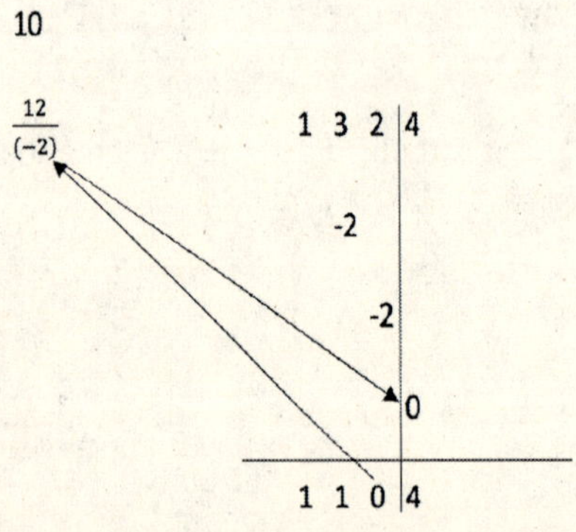

Answer: 1324 divided by 12 gives 110 as quotient and 4 as remainder.

Verification: As discussed in the last chapter, we can verify our result by using the rule:

$$dividend = quotient \times divisor + remainder$$

Applying this rule, we get:

$$dividend = 110 \times 12 + 4$$

$$dividend = 1320 + 4$$

$$dividend = 1324$$

Hence verified.

Example 2: 4752 ÷11

Step 1: Selected base is 10 and the difference will be (-1).

Step 2: Since the base has only 1 zero, the dividend will be split up in such a way that the RHS will have only one digit.

10

$\frac{11}{(-1)}$ 4 7 5 | 2

Quotient | Remainder

Step 3: Bring down the first digit of the dividend, 4, as it is. This is the first digit of the quotient.

10

$\frac{11}{(-1)}$ 4 7 5 | 2

4 |

Step 4: Multiply this digit with the difference and write the product under the second digit of the dividend. In this case, it will be 4 x (-1) = -4. Add the digits in the second column of the dividend to get the second digit of the quotient.

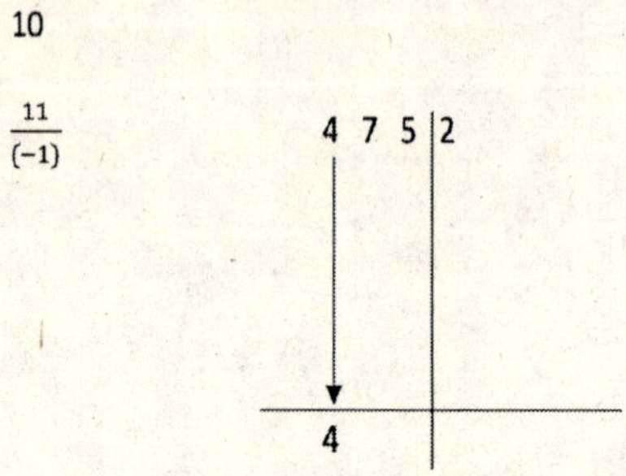

Step 5: multiply the second digit of the quotient with the difference and write the product under the next digit, 5, of the dividend. Add the third column to get the third digit of the quotient.

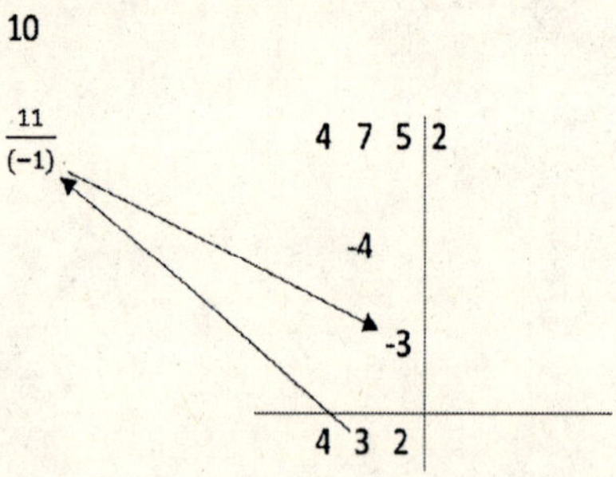

Step 6: Multiply third digit of the quotient with the difference and write the product under the next digit, 2, of the dividend. Add the digits in this column to get the remainder. 10

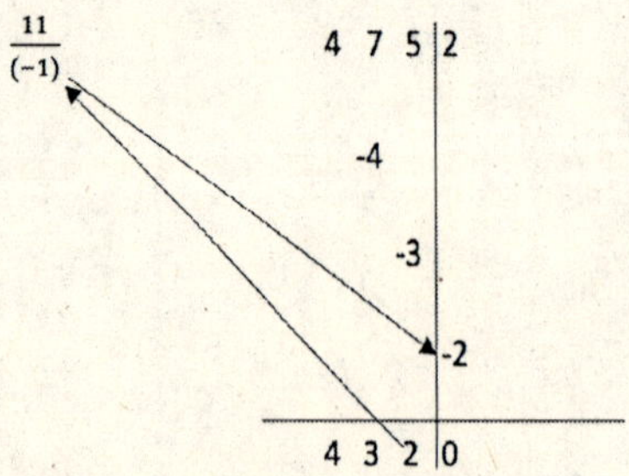

Answer: 4752 ÷11 gives 432 as quotient and 0 as remainder.

Example 3: 1328 ÷ 13

Step 1: Base: 10; Difference: (-3)

Step 2: Splitting the dividend

10

$\frac{13}{(-3)}$

1 3 2	8
Quotient	Remainder

Step 3:

10

$\frac{13}{(-3)}$

1 3 2	8
1	

Step 4:

10

$\frac{13}{(-3)}$

1 3 2	8
-3	
1 0	

Step 5:

10				
$\frac{13}{(-3)}$	1	3	2	8
		-3	0	
	1	0	2	

Step 6:

10				
$\frac{13}{(-3)}$	1	3	2	8
		-3	0	
				-6
	1	0	2	2

Answer: 1328 ÷ 13 gives 102 as quotient and 2 as remainder.

Since the difference in this case is a negative number, it is possible that while adding a column, the answer comes as a negative number. Let us see what steps to take when that happens.

Example 4: 4681 ÷ 11

Applying the above-mentioned method, we get:

10				
$\frac{11}{(-1)}$	4	6	8	1
		-4		
			-2	
				-6
	4	2	6	-5

As we can see, in the remainder column, we have a negative number (-5). Since the remainder cannot be negative, we carry 1 from the quotient to the remainder side, which means adding the divisor once to the remainder, that is, adding 11 to (-5). The answer we get is 6 and we need to remember that 1 has been deducted from the quotient. Hence, our final answer is:

4681÷11 gives 425 as quotient and 6 as remainder.

We can verify this using the verification rule:

$$dividend = quotient \times divisor + remainder$$

Dividend = 425 x 11 + 6

= 4675 + 6

= 4681. Hence verified.

8.2 When the base is 100

Example 5: 16453 ÷ 102

Step 1: Since 102 is closer to 100, we will select 100 as the base and the difference between the base and the divisor will be (-02). The difference will have 2 digits as the selected base has 2 zeros. Also, the product at each stage will also have 2 digits.

Step 2: Splitting the dividend: The dividend will be split in such a way that the RHS will have 2 digits as the number of zeros in the base is 2.

100

$\frac{102}{(-02)}$	1 6 4	5 3
	Quotient	Remainder

Step 3: Bring down the first digit, 1, as it is. This is the first digit of the quotient.

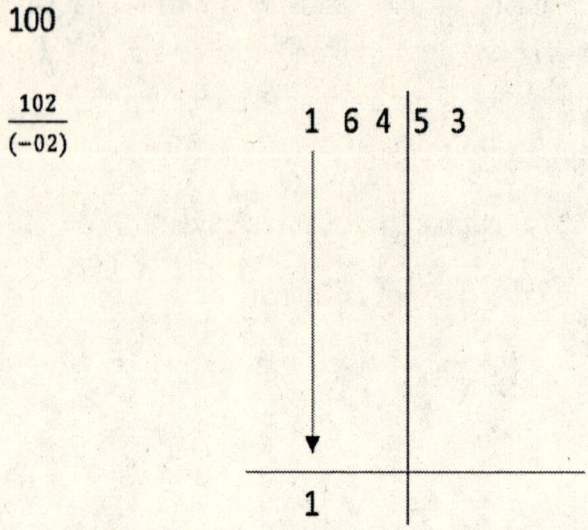

Step 4: Multiply this digit with the difference and write the product under the next two digits of the dividend. The product in this case will be 1 x (-02) = -02 and this will be written under 6 and 4. Since there are two digits in the product and the product is a negative number, we will put a bar over the numbers to denote that they are negative numbers. This will help us in recognizing negative numbers when we are adding numbers in a column. So, we will write (-02) as $\overline{0}\,\overline{2}$.

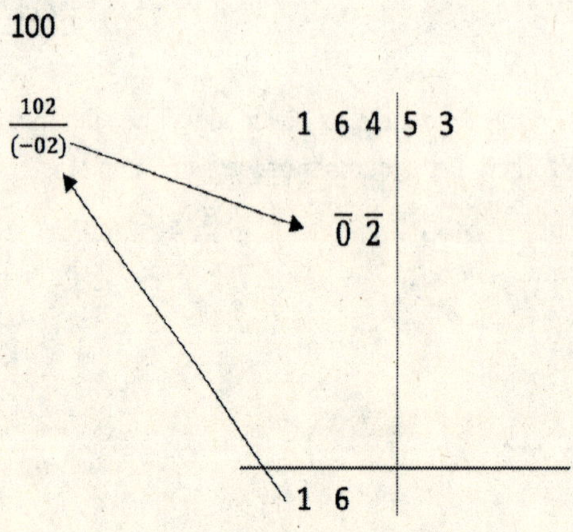

Step 5: Multiply this digit, 6, with the difference and write the product under the next two digits of the dividend. In this case, it will be 6 x (-02) = -12. This -12 will be written as $\bar{1}\,\bar{2}$, under digits 4 and 5 of the dividend. Adding the numbers in the third column we get, 4 + (-2) + (-1) = 1.$\frac{102}{(-02)}$

$$\frac{102}{(-02)} \qquad \begin{array}{ccc|cc} 1 & 6 & 4 & 5 & 3 \\ & \bar{0} & \bar{2} & & \\ & & \bar{1} & \bar{2} & \\ \hline 1 & 6 & 1 & & \end{array}$$

Step 6: Multiply the third digit of the quotient with the difference and write the product under the next two digits, 5 and 3, of the dividend.

1 x (-02) = (-02). This will be written as $\bar{0}\,\bar{2}$ under the digits 5 and 3. Adding the digits in the remainder column, we get 5 + (-2) + (0) = 3 and 3 + (-2) = 1

$$\frac{102}{(-02)} \qquad \begin{array}{ccc|cc} 1 & 6 & 4 & 5 & 3 \\ & \bar{0} & \bar{2} & & \\ & & \bar{1} & \bar{2} & \\ & & & \bar{0} & \bar{2} \\ \hline 1 & 6 & 1 & 3 & 1 \end{array}$$

Final answer: 16453 ÷ 102 gives 161 as quotient and 31 as remainder.

Example 6: 18754 ÷ 103

Step 1: Base: 100; difference is (-03)

Step 2: Splitting the dividend.

100

$\frac{103}{-03}$

1	8	7	5	4
Quotient			Remainder	

Step 3:

$\frac{103}{-03}$

1	8	7	5	4
↓				
1				

Step 4:

100

$\frac{103}{-03}$

1	8	7	5	4
	$\bar{0}$	$\bar{3}$		
1	8			

Step 5:

$\frac{103}{-03}$

1	8	7	5	4
	$\bar{0}$	$\bar{3}$		
		$\bar{2}$	$\bar{4}$	
1	8	2		

Step 6:

$$\frac{103}{-03}$$

$$\begin{array}{ccc|cc}
1 & 8 & 7 & 5 & 4 \\
 & \overline{0} & \overline{3} & & \\
 & & \overline{2} & \overline{4} & \\
 & & & \overline{0} & \overline{6} \\
\hline
1 & 8 & 2 & 0 & 8
\end{array}$$

Note: when we are adding the digits in the remainder column, we can treat both the columns at the same time and do the addition. Like in this case, it is 54 + (-40) + (-06) = 08.

Example 7: 10121 ÷ 112

$$\frac{112}{-12}$$

$$\begin{array}{ccc|cc}
1 & 0 & 1 & 2 & 1 \\
 & \overline{1} & \overline{2} & & \\
 & & 1 & 2 & \\
 & & & 0 & 0 \\
\hline
1 & \overline{1} & 0 & 4 & 1
\end{array}$$

Note: In this case, when we bring down the second quotient digit, it is (-1). Now when this (-1) is multiplied with the difference we get (-1) x (-12) = 12. In the next step we get 00 as 0 is sum of digits in the third column.

When we have a negative in the quotient, like in this case, then it is written as: as the negative part is 10

$$1\overline{1}0 = 100 - 10 = 90$$

Hence the quotient is 90 and the remainder is 41.

Example 8: 21141 ÷ 107

$$\frac{107}{-07}$$

$$\begin{array}{ccc|cc}
2 & 1 & 1 & 4 & 1 \\
 & \overline{1} & \overline{4} & & \\
 & & 0 & 0 & \\
 & & & 2 & 1 \\
\hline
2 & 0 & \overline{3} & 6 & 2
\end{array}$$

In this question also, we get a negative number as the sum of the digits in the third column. When this negative number is multiplied with the negative difference, the answer is positive. (-3) x (-07) = 21

But in the quotient, we have a negative digit. To get the actual quotient, we subtract the negative part from the base / multiple of base:

$20\overline{3}$ = 200 – 3 =197

Hence the quotient is 197 and the remainder is 62.

8.3 When the base is 1000

Example 9: 21432 ÷ 1002

Step 1: Here the divisor is 1002 which is close to 1000. So, the selected base will be 1000 and the difference will be (-002). Note that the difference will be written as three digits as the base has three zeros. When we write the product of the quotient digit and the difference, that will also be written as three digits.

Step 2: The dividend will be split in such a way that the RHS has three digits.

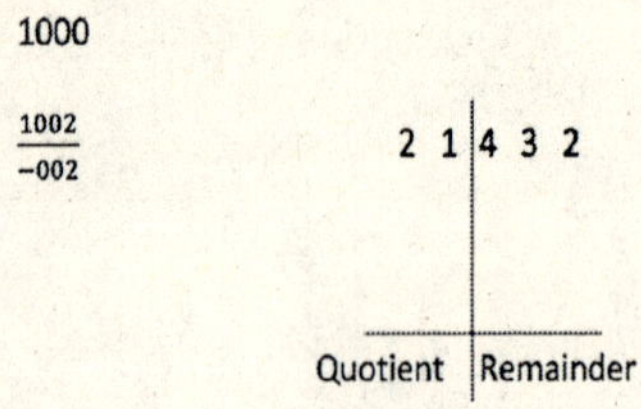

Step 3: Bring down the first digit of the dividend as it is. This becomes the first digit of the quotient.

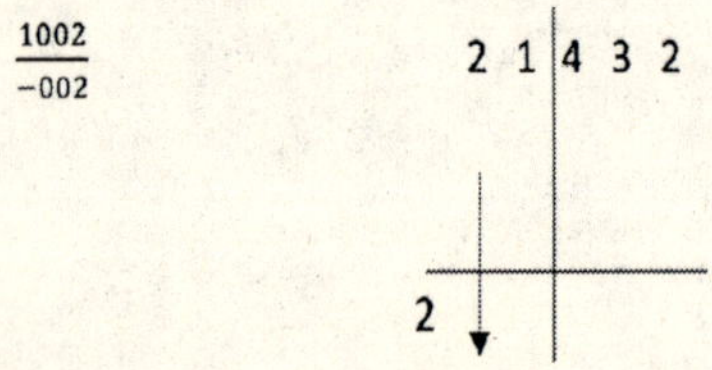

Step 4: Multiply this digit, 2, with the difference and write the product under the next three digits of the dividend. Since the product will be negative [2 x (-002) = -004], it will be written as a number with a bar.

$$\frac{1002}{-002} \qquad \begin{array}{cc|ccc} 2 & 1 & 4 & 3 & 2 \\ & \bar{0} & \bar{0} & \bar{4} & \\ \hline 2 & & & & \end{array}$$

Step 5: Add the digits in the second column of the dividend, to get the second digit of the quotient. It will be (1 – 0 = 1).

Multiply this digit with the difference and write the product under the next three digits of the dividend.

$$\frac{1002}{-002} \qquad \begin{array}{cc|ccc} 2 & 1 & 4 & 3 & 2 \\ & \bar{0} & \bar{0} & \bar{4} & \\ & & \bar{0} & \bar{0} & \bar{2} \\ \hline 2 & 1 & 3 & 9 & 0 \end{array}$$

Step 6: Adding the digits in the remainder columns, we get the 432 + (-040) + (-002) = 390

Answer: 21432 ÷1002 gives 21 as quotient and 390 as remainder.

Example 10: 17893 ÷ 1005

$$\frac{1005}{-005} \qquad \begin{array}{cc|ccc} 1 & 7 & 8 & 9 & 3 \\ & \bar{0} & \bar{0} & \bar{5} & \\ & & \bar{0} & \bar{3} & \bar{5} \\ \hline 1 & 7 & 8 & 0 & 8 \end{array}$$

Answer: 17893 ÷ 1005 gives 11 as quotient and 838 as remainder.

Example 11: 223987 ÷1003

Divisor						
1003 / −003	2	2	3	9	8	7
		$\bar{0}$	$\bar{0}$	$\bar{6}$		
			$\bar{0}$	$\bar{0}$	$\bar{6}$	
				$\bar{0}$	$\bar{0}$	9
	2	2	3	3	1	8

Answer: 223987 ÷ 1003 gives 223 as quotient and 318 as remainder.

Summary:

- Select the base and find the difference between the base and the number. If this difference is negative, write it as a negative number.
- Split the dividend into two parts such that the RHS has the same number of digits as the zeroes in the base.
- Bring down the first digit of the LHS of the dividend as it is. This becomes the first digit of the quotient.
- Multiply this digit with the difference and write the product under the next digits of the dividend. Note that the product will have as many digits as the zeroes in the base and if the difference is negative, the product will also be negative.
- Add the digits in the second column of the dividend. Be mindful of the negative sign. This sum becomes the second digit of the quotient.
- This process of step 4 and 5 is repeated till we have written a digit under the last column of the dividend.
- The LHS of the answer gives us the quotient and the RHS gives us the remainder.
- If the remainder is bigger than the divisor, then we divide the remainder with the divisor again and the quotient thus obtained is added to the quotient obtained using base division and the remainder obtained will be the final remainder.

Try it yourself

Q1. Find the quotient and the remainder:

a) 435 ÷ 13
b) 321 ÷ 12
c) 6541 ÷ 11
d) 4781 ÷ 14
e) 2782 ÷ 12

Q2. Find the quotient and the remainder:

a) 1435 ÷ 103
b) 3027 ÷ 102
c) 16541 ÷ 111
d) 24981 ÷ 104
e) 22782 ÷ 102

Q3. Find the quotient and the remainder:

a) 44935 ÷ 1003
b) 19321 ÷ 1002
c) 116501 ÷ 1001
d) 284789 ÷ 1004
e) 132782 ÷ 1000

Chapter 9

General division

In the previous two chapters we have seen how the techniques of Vedic mathematics can simplify the division process if the divisor is near the base numbers. In this chapter we will learn about long division by any divisor.

The conventional method of long division is as shown below. In this method, the student makes an approximate guess for the first digit of the quotient, multiplies it with the divisor and writes it down below the dividend and computes the remainder. Then the process is repeated when the next digit of the dividend is written along with the remainder of the first step. This is a time consuming and a tedious process which requires a student to do multiplication of large numbers and computing remainders at every step.

```
43)74975(1743
43
----
319
301
 ----
 187
  172
  ----
  155
  129
   ----
   26
```

Answer: 74985 ÷ 43 gives 1743 as quotient and 36 as remainder.

In the Vedic mathematics method, no matter how long the divisor is, the dividend will be divided only by the first digit of the divisor and yet the answer will be accurate. We will be considering the digits of the divisor as two separate digits(in case of a two digit divisor), but we will perform the division only with the first digit of the divisor.

9.1 When the divisor is a two-digit number.

Let us understand this method with the help of the same example.

Example: 75985 ÷ 43

In this case, our divisor is 43. We will treat 4 and 3 as two separate digits but we will be performing the division with 4. We will set up the question like a simple division question as shown below:

```
              Quotient

D1 | D2 | Dividend
        |

         __________
          Remainder
```

Here D1 and D2 denote the two digits of the divisor. For example, in this question D1 is 4 and D2 is 3.

```
            Quotient

4 | 3    | 7  4 9 7 5
         |

          ___________
           Remainder
```

Step 1: Divide the first digit of the dividend, 7, by D1, 4. We will get 1 as quotient. Write this 1 as the first digit of the quotient, Q1.

```
           1 ←—— Q1

4 | 3    | 7  5 9 7 5
         |
```

Step 2: Multiply Q1 by D1 and write the product under the first digit of the dividend and multiply Q1 by D2 and write the product under the second digit of the dividend as shown.

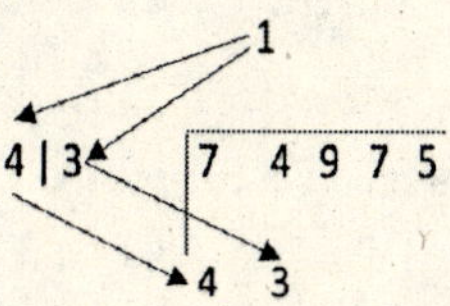

Step 3: Subtraction: The subtraction in this case is slightly different from the normal subtraction. We perform the subtraction in the first column of the dividend, to get, 7 – 4 = 3. This 3 is written alongside the second digit of the dividend, making it 34. Then the subtraction in the second column is performed giving us 34 – 3 = 31.

```
              1
4 | 3    | 7  34  9 7 5
         | 4   3
         ----------
           3  31
```

Step 4: Division: Now our working dividend is 31. We divide 31 by D1, 4, to get the second digit of the quotient as 7. This is Q2.

```
              1  7 ← Q2
4 | 3    | 7  34  9 7 5
         | 4   3
         ----------
              31
```

Step 5: Multiply the digit Q2, 2, with D1 and write the product under 31 and multiply Q2 with D2 and write the product under the next digit of the dividend, 9.

```
              1  7 ← Q2
4 | 3    | 7  34  9 7 5
         | 4   3  21
         ----------
              31
              28
```

Step 6: Perform the subtraction in the second column (31 – 28 = 3). Write this digit, 3, along with the next digit in the dividend, which is 9 and make it 39. Now perform the subtraction in the third column of the dividend. (39 – 21 = 18). 18 is our next working dividend.

```
              1  7 ←— Q2
4 | 3     | 7  34  397  5
          4  3  | 21
             31 | 18
             28
             3
```

Step 7: Divide 18 by D1, 4, to get the next digit of the quotient as 4. This is Q3.

```
              1  7  4   ←— Q3
4 | 3     | 7  34  397  5
          4  3  21
             31  18
             28
```

Step 8: Multiply Q3 with D1 and write the product under 18. Multiply Q3 with D2 and write the product under the next digit, 8, of the dividend. 4 x 4 = 16 and 4 x 3 = 12

16 will be written under 18 and 12 will be written under 8 as shown below:

```
              1  7  4
4 | 3     | 7  34  39   7   5
          4  3  21   12
             31  18
             28  16
```

Step 9: Perform the subtraction in the third column of the dividend

18 -16 = 2. Write this 2 along with the next digit of the dividend, 8 and make it 28. Perform the next subtraction 28 – 12 = 16. 16 is now our working dividend.

```
              1  7   4

4 | 3      | 7  34  39  27  5

             4   3  21 | 12
           ____________|
                31  18 | 15

                28  16 |
              ________ |
                    2
```

Step 10: We divide the working quotient, 15 by D1, 4, to get the next digit of the quotient.

15 ÷ 4 gives 4 as quotient. This is Q4.

Multiply Q4 by D1 and write the product under 16 and multiply Q4 by 3 and write the product under the next digit of the dividend, which is 5.

4 x 3 = 12 and 3 x 3 = 9

```
              1  7   4  3 ←—— Q4

4 | 3      | 7  34  39  27  5

             4   3  21  12  9
           ____________
                31  18  15

                28  16  12
              ________
```

Step 11: Perform the next subtraction, 15 -12 = 3. Write this 3 along with the next digit of the dividend, 5 and make it 35. Perform the subtraction in the last column to give 35 – 9 = 26. Since we have exhausted all the digits of the dividend, this is our final remainder.

		1	7	4 3 ← Quotient		
4 | 3	7	34	39	27	35	
	4	3	21	12	9	
		31	18	15	26 ← Remainder	
		28	16	12		
				3		

Answer: Quotient is 1743 and the remainder is 26.

At first glance, this method appears to have many more steps than the conventional method but if you look carefully, there are no complex multiplications. With practice the entire sum can be done in 2 steps. Let us look at a few more examples.

Example 2: 85216 ÷ 72

Step 1: Set up the problem by treating the divisor as two individual digits. D1 is 7 and D2 is 2.

7 | 2	8	5	2	1	6

Step 2: Divide the first digit of the dividend, 8, by D1. The quotient of this division will be the first digit of the required quotient. In this case 8 divided by 7 gives 1 as quotient.

Step 3: Multiply this digit, 1, by D1 and write the product under the first digit of the dividend. Multiply this digit by D2 and write the product under the second digit of the dividend.

1 x 7 = 7; 1 x2 = 2

		1 ← Q1			
7 | 2	8	5	2	1	6
	7	2			

Step 4: Do the subtraction in the first column of the dividend. Write the difference along with the second digit of the dividend before proceeding with the subtraction of the second column. 8 – 7 gives 1. This 1 is written along with 5 to give 15. Then second column subtraction is 15 – 2 = 13. This is our working dividend for the next step.

```
            1 ←—— Q1
7 | 2   | 8  15  2  1  6
        | 7  2
        ----------------
          1  13
```

Step 5: Divide the working dividend with D1. The quotient of this division will give us the second digit of our required division. 13 divided by 7 will give us 1 as quotient.

Step 6: Multiply this digit, 1, by D! and write the product under the working dividend, 13. Multiply this digit by D2 and write the product under the next digit 2 of the dividend.

```
            1 1
7 | 2   | 8  15  2  1  6
        | 7  2   2
        ----------------
          1  13
             7
```

Step 7: Do the subtraction 13 – 7 = 6 and write the difference along with the next digit of the dividend and then perform the subtraction 62 – 2 = 60.

60 is our working dividend now.

```
            1 1
7 | 2   | 8  15  62 1  6
        | 7  2 | 2
        ----------------
          1  13| 60
             7 |
             6
```

Step 8: Divide the working dividend, 60, with D1 to get the next digit of the quotient as 8.

Step 9: Multiply 8 with D1 and write the product under the working dividend. Multiply 8 with D2 and write the product under the next digit of the dividend, which is 1.

8 x 7 = 56; 8 x 2 = 16

```
          1 1 8
7 | 2    |8 15 62 1  6
         |7 2   2 16
         ----------------
          1 13 60
             7 56
         ----------------
             6
```

Step 10: Perform the subtraction in the third column of the dividend, 60 – 56 = 4. Write the difference with the next digit of the dividend, 1, to make it 41 and then perform the subtraction in that column, 41 – 16 = 25. The working dividend is now 25.

Step 11: Divide the working dividend, 25, with D1 to get the next digit of our required quotient. 25 divided by 7 gives us the quotient 3. This is the next digit of the required quotient.

Step 12: Multiply 3 by D1 and write the product under the working dividend 25. Multiply 3 with D2 and write the product under the next digit of the quotient which is 6.

3 x 7 = 28; 3 x 2 = 6

```
          1 1 8 3
7 | 2    |8 15 62 41 6
         |7 2   2 16 6
         ----------------
          1 13 60 25
             7 56 21
         ----------------
             6  4
```

Step 13: Perform the subtraction 25 – 21 = 4 and write this difference with the next digit of the dividend, 6, to make it 46 and then perform the subtraction in that column to give 46 – 6 = 40

```
            1 1  8  3
7 | 2    | 8 15 62 41 46
         | 7 2    2 16 | 6
         ---------------------
           1 13 60 25 | 40
               7 56 21
         ---------------------
               6  4  4
```

Since we have performed the subtraction in the last column, we stop at this and we get our final answer. **The quotient is 1183 and the remainder is 40.**

Example 3: 68705 ÷ 54

Step 1:

```
5 | 4  |  6  8  7  0  5
```

Step 2:

```
           1
5 | 4  |  6  18  7  0  5
          5 | 4
         ------------------
          1   14
```

Step 3:

```
            1   2
5 | 4  |  6   18  47  0  5
          5    4 | 8
         --------------------
          1   14 | 39
              10
         --------------------
               4
```

Step 4:

```
             1   2   7

  5 | 4  |  6   18  47  40  5
            5    4   8 | 28
          ----------------------
            1   14  39 | 12
                10  35 |
              ----------------------
                 4   4
```

Step 5:

```
        1   2   7   2   <------  Quotient

  5 | 4  |  6   18  47  40  25
            5    4   8  28 |  8
          ---------------------
            1   14  39  12 | 17  <------  Remainder
                10  35  10 |
              ---------------------
                 4   4   2
```

Answer: 68705 ÷ 54 gives 1272 as quotient and 17 as remainder.

Example 4: 78321÷ 63

Step 1:

```
  6|3  |  7   8   3   2   1
```

Step 2:

```
          1
  6|3  |  7  18   3   2   1
          6   3
        -----------------------
          1  15
```

Step 2:

	1	2			
6|3	7	18	33	2	1
	6	3	**6**		
	1	15	**27**		
		12			
		3			

Step 3:

	1	2	4		
6|3	7	18	33	32	1
	6	3	6	**12**	
	1	15	27	**20**	
		12	**24**		
		3	**3**		

Step 4:

	1	2	4	3 ← Quotient		
6|3	7	18	33	32	21	
	6	3	6	12	**9**	
	1	15	27	20	**12**	← Remainder
		12	24	**18**		
		3	3	**2**		

Answer: 78321 ÷ 63 gives 1243 as quotient and 12 as remainder.

Example 5: 92235 ÷ 41

	2	2	4	8	Quotient	
4|1	9	12	21	37	15	
	8	2	2	4	8	
	1	10	19	33	7	← Remainder
		8	16	32		
		2	3	1		

As we have seen in the above examples, Vedic math division does not involve guess work and approximation and repeated multiplication to get the answer. It simplifies the division by splitting the divisor. However, it may happen that subtraction in one of the steps gives us a negative answer or the working dividend may be 0. Let us see how we solve in that case.

Example 6: When the subtraction in one of the columns is negative.

8754 ÷ 69

Solution:

	1	~~3~~2	7	
6 | 9	8	27	69	54
	6	9	18	63
	2	18	51	
		12	42	
		6	5	

In this case, we divide the first digit of the dividend, 8, by D1 of the divisor, 6. This gives us the first digit of the quotient, 1. Multiply 1 by both D1 and D2 and write the product under the first and the second digit respectively. Perform the subtraction in the first column. The difference from the first column is written along with the second digit of the dividend to make it 27. Then the subtraction of the second column is carried out. The working dividend comes out to be 18.

Dividing the working dividend by D1 gives us the second digit of the quotient as 3. Multiply the second digit of the quotient with D1 and D2 and write the product under the working dividend and the third digit of the dividend respectively. Performing the subtraction in the second column gives us 0. Writing this 0 next to the third digit gives us 05 and it is not possible for us to subtract 27 from 5. The solution lies in going back one step and reducing the quotient by 1. In this case we will make the second digit of the quotient as 2 and repeat the last step with 2 as the quotient.

In this step also the last division (54 – 63) will give us a negative answer. So we reduce the previous quotient by 1 and make it 6.

```
           1  32  76
6 | 9  |  8  27 |65 |114
          6   9 |18 |54
       -----------------
          2  18 |47 |60  <------ Remainder
             12 |36
       -----------------
              6  11
```

Answer: 8754 ÷ 69 gives 126 as quotient and 60 as remainder.

Example 7: When the working dividend is 0

7741 ÷ 43

Solution:

```
          1  8
4 | 3  |  7 37  24  1
          4| 3 |24
       ----------------
          3 34 | 0
            32 |
       ----------------
             2
```

Since at this stage we have the working dividend as 0, we will write 0 as the next quotient digit and multiply D1 and D2 by 0 and write the products under the working dividend and the next digit respectively.

```
          1  8   0
4 | 3  |  7 37  24  1
          4| 3 |24  0
       ----------------
          3 34 |  0  1
            32 |  0
       ----------------
             2    0
```

Answer: 7741 ÷ 43 gives 180 as quotient and 1 as remainder.

9.2 When the divisor is a three-digit number

When we divide by a three-digit number, the process is nearly the same as the two-digit number. Here we will split the divisor into three separate digits, D1, D2 and D3. Let us understand it with the help of an example.

Example 8: 75243 ÷ 241

Step 1: Split the divisor into three separate digits and set up the problem.

In this case 2 is D1, 4 is D2 and 1 is D3.

```
2 | 4 | 1 | 7  5  2  4  3
```

Step 2: Divide the first digit of the dividend, 7, by D1. The quotient of this division is the first digit of our required division. In this case 7 ÷ 2 gives quotient as 3. This is written as the first digit of the quotient, Q1. Multiply Q1 with D1, D2 and D3 separately and write the products under the first, second and third digit of the dividend respectively.

3 x 2 =6; 3 x 4 = 12; 3 x 1 = 3

```
            3
2 | 4 | 1 | 7  5  2  4  3
            6  12 3
```

Step 3: Perform the subtraction in the first column of the dividend. 7 – 6 = 1. Write this difference, 1, along with the second digit of the dividend, making it 15. After writing the 1, carry out the subtraction in the second column of the dividend. 15 – 12 = 3. This difference is written along with the third digit of the dividend, making it 32. Now carry out the subtraction in the third column of the dividend, 32 – 3 = 29. This 29 is our working dividend.

This is very similar to the division by two-digit divisor but here the difference is written next to the digit twice and the subtraction is performed three times in a single step.

Another difference is that the digits of the working dividend will be treated as two separate digits. We will see how that works in the subsequent steps.

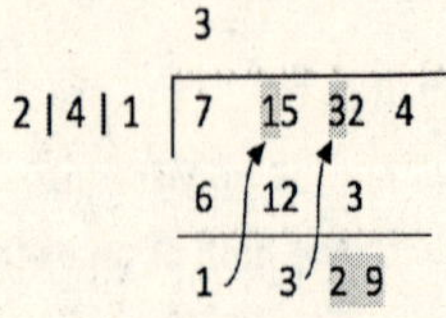

Step 4: Now consider the first digit of the working dividend, which is 2 in this case and divide it by D1. The quotient of this division becomes the second digit of our required quotient.

2 ÷ 2 gives 1 as quotient. So, 1 is the second digit of the required quotient. 1 is then multiplied by D1, D2 and D3 and the products written under the first digit of the working dividend, second digit of the working dividend and the next digit of the dividend respectively.

1 x 2 = 2 (written under 2); 1 x 4 = 4 (written under 9); and 1 x 1 = 1 (written under 4)

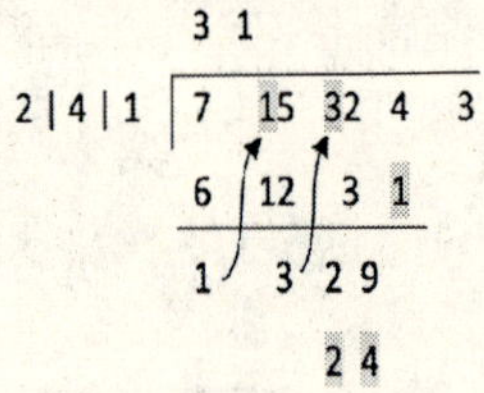

Step 5: Perform the subtraction under the first digit of the working dividend; 2 – 2 = 0. Since it is 0, we need not write it next to 9. But had it been any other digit, it would have been written along with 9 and then subtraction under that column would have been carried out. Here we directly do 9 – 4 = 5. We write this difference along with the next digit of the dividend, 4, making it 54 and then carry out the subtraction in that column, 54 – 1 = 53. Now 53 is our working dividend.

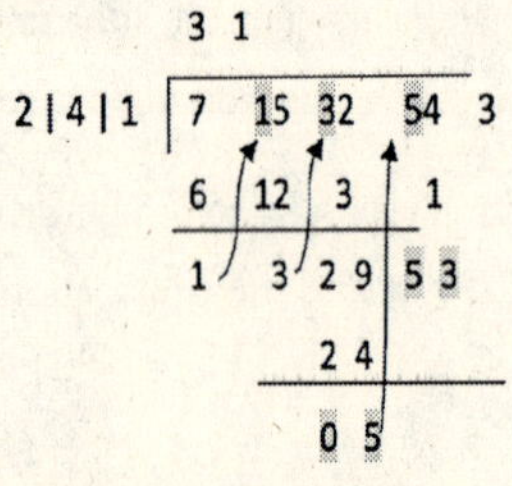

Step 6: Repeat step 4 and 5 for the working dividend as 53. We will treat 53 as two separate digits, 5 and 3 and repeat the steps for finding the third digits of the quotient, then multiplying it with D1, D2 and D3 of the divisor and performing the subtraction as shown.

5 ÷ 2 gives 2 as quotient. This is the third digit of our required quotient.

2 x 2 = 4 (written under 5); 2 x 4 = 8 (written under 9); 2 x 1 = 2 (written under 3);

Performing the column wise subtraction, we get,

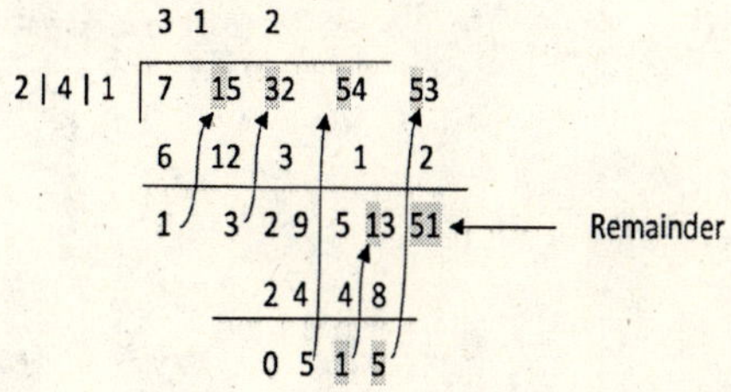

Answer: 75243 ÷ 241 gives 312 as quotient and 51 as remainder.

Example 9: 79342 ÷ 562

Step 1:

5 | 6 | 2 | 7 9 3 4 2

Step 2:

```
          1
5 | 6 | 2 | 7  9  3  4  2

            5  6  2
```

7 divided by 5 gives quotient 1. This 1 is multiplied by 5, 6 and 2 separately and the products written under 7, 9 and 3 respectively.

Step 3:

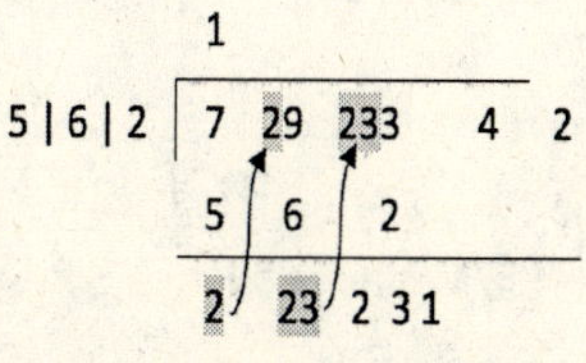

7 – 5 = 2. 2 is written next to 9 to make it 29.
29 – 6 = 23. 23 is written next to 3 to make it 233.
233 – 2 = 231.
231 is the working dividend.

Step 4:

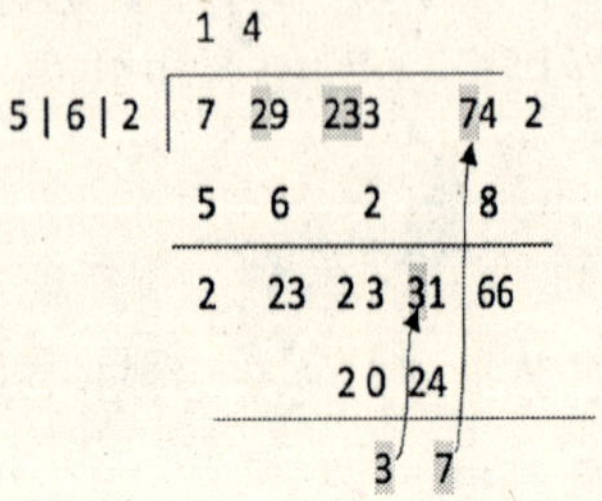

Since the first digit of the working dividend is smaller than D1, take the first two digits of the working dividend together. 23 divided by 5 gives 4 as quotient.

4 x 5 = 20; 4 x 6 = 24; 4x 2 = 8. These are written under 23, 1 and 4.

Subtraction: 23 – 20 = 3. 3 is written next to 1 making it 31.

31 – 24 = 7. 7 is written next to 4 making it 74.

74 – 8 = 66. 66 is the working dividend.

Step 5:

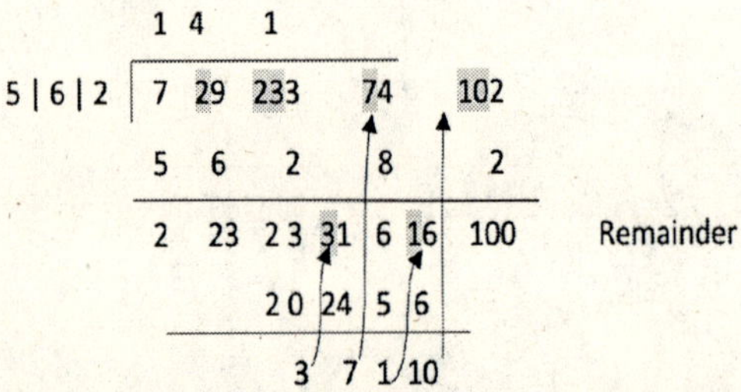

Answer: 79342 ÷ 562 gives 141 as quotient and 100 as remainder.

Example 10: 25711 ÷ 724

Since the first digit of the dividend is smaller than the first digit of the divisor, we will take the first two digits together for the first step.

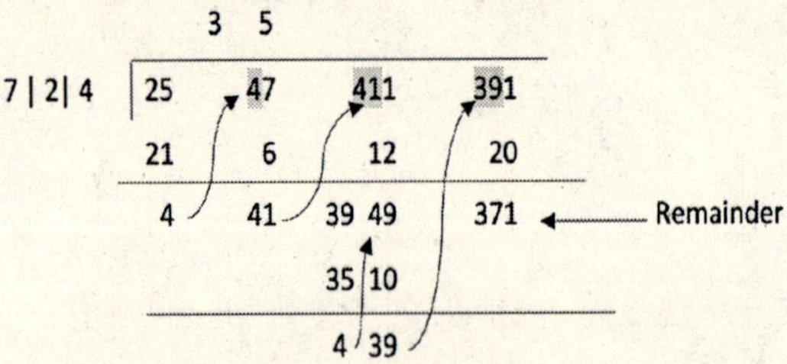

Answer: 25711 ÷ 724 gives 35 as quotient and 371 as remainder.

Example 11: 74291 ÷ 347

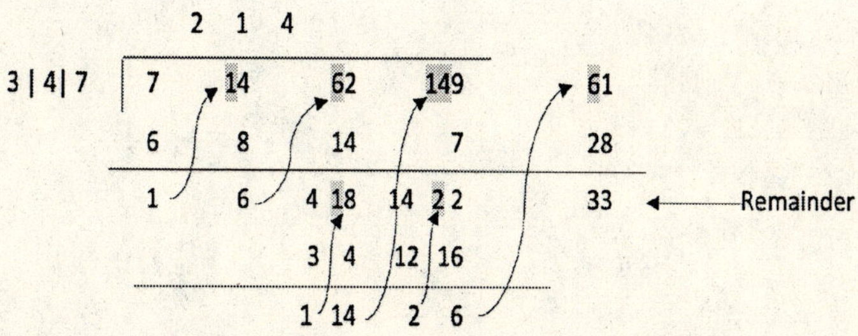

Answer: 74291 ÷ 347 gives 214 as quotient and 33 as remainder.

From these examples, it is clear that long division by Vedic mathematics reduces the complexities of the conventional long division method and with practice, it can be done in two or three steps with simple single digit multiplication and subtraction.

Challenge:

Can you try and write the steps for division by a four-digit divisor?

Try it yourself

Q1. Find the quotient and the remainder:

a) 4356 ÷ 72
b) 6752 ÷ 42
c) 93421 ÷ 48
d) 4781 ÷ 68
e) 7821 ÷ 32
f) 24356 ÷ 17
g) 68923 ÷ 28
h) 5732 ÷ 29
i) 95432 ÷ 31
j) 76321 ÷ 62

Q2. Find the quotient and the remainder:

a) $345212 \div 214$
b) $67542 \div 114$
c) $783212 \div 231$
d) $85642 \div 371$
e) $976521 \div 421$

Chapter 10

Squares

Squaring a number means multiplying the number by itself. For example, square of 4 means 4 x 4 = 16. So, we say 16 is the square of 4. Mathematically we write it as:

$$4^2 = 16$$

Similarly, $5^2 = 5 \times 5 = 25$ and

$12^2 = 12 \times 12 = 144$

We can find the squares of all real numbers by multiplying the number by itself. Note that the squares of all real numbers are positive as multiplying a negative number by itself will also yield a positive answer.

$$(-2)^2 = (-2) \times (-2) = 4$$

Square root is the reverse operation of squaring. It is denoted by the symbol " "$\sqrt{\square}$". For example, if the square of 4 is 16, then we say square root of 16 is 4. However, we must remember that square of (-4) will also be 16. But while writing the square root we write only the positive square root of the number.

$$\sqrt{16} = 4$$

In conventional method, there is only one way of finding the square of a number, namely, multiplying the number by itself using conventional multiplication. Vedic mathematics, on the other hand, provides for interesting and easy ways to find squares of numbers. In this chapter we will considering some specific cases and then discuss a general method of finding squares, which can be applied to any number.

10.1 Squares of numbers ending in 5.

Numbers ending in 5 implies numbers that have 5 in their unit's place, like 15, 25, 75, 85 etc. In Vedic mathematics, the method for finding the squares of such numbers follows the rule "***by one more than the one before***".

Steps for finding the square of a number ending in 5:

1. The answer comes in two parts – LHS and RHS.
2. RHS is always $5^2 = 25$.
3. LHS is computed by multiplying the digit other than 5, by its successor, that is by one more.

Example 1: Find the square of 75.

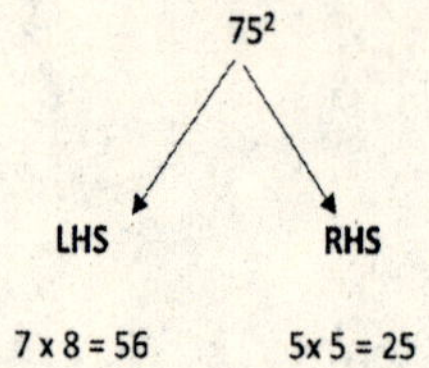

The digit other than 5 is 7. The successor of 7 is 8. Hence, LHS is 7 x 8 = 56

RHS is simply the square of 5 which is 25.

Combining the LHS and RHS we get the answer as **75^2 = 5625.**

Example 2: Find the square of 45

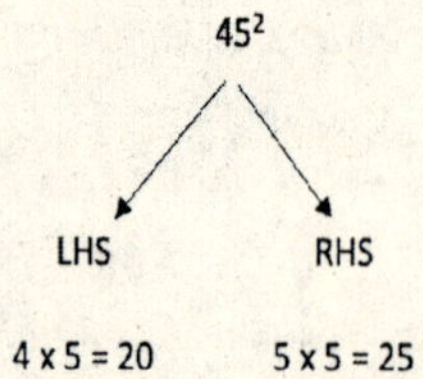

Answer: 2025

Example 3: Find the square of 85

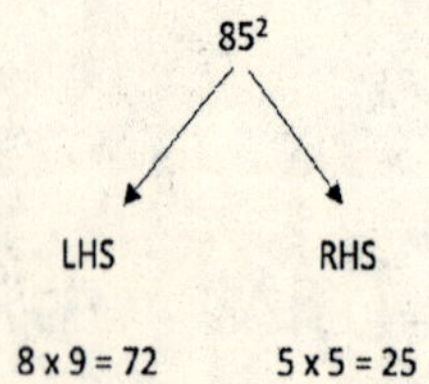

Answer: 7225

Example 4: Find the square of 115

115^2

LHS RHS

11 x 12 = 132 5 x 5 = 25

Answer: 13225

Note that whatever digits are present in the number excluding 5, they are considered as a single number. For example, in this case, the number was 115. Excluding 5, we get 11. So, it is multiplied by its successor 12 to get the LHS.

Another interesting multiplication:

The above method is not only applicable to squaring of numbers ending in 5 but also to the multiplication of numbers whose last digits add up to 10 and the remaining digits are same.

Thus, there are two conditions necessary for this multiplication:

1. The digits in the unit's place must add up to 10
2. Remaining digits should be same

Example 5: Find the product 72 x 78.

Here we see that the digits in the unit's place add up to 10 (2 + 8 = 10) and the remaining digit is 7 in both the numbers. Hence, we can apply this technique.

Steps:

1. To get the LHS, multiply the digit which is same by its successor

7 x 8 = 56. 56 is the LHS of the required answer.

2. To get the RHS, find the product of the digits in the unit's place. RHS must be a two-digit answer.

2 x 8 = 16

Answer: 5616

Example 6: Find the product 91 x 99

Step 1: LHS: 9 x 10 = 90

Step 2: RHS: 1 x 9 = 9 = 09

Answer: 9009

Example 7: 64 x 66

Step 1: LHS: 6 x 7 = 42

Step 2: RHS: 4 x 6 = 24

Answer 4224

10.2 Squares of numbers between 40 and 60.

The numbers between 40 and 60 are special cases as their squares can be easily found using the square of 50 which is 2500.

Steps:

1. The answer will have two parts – LHS and RHS
2. For computing the LHS:
    (a) Find the difference between the number and 50. If the number is between 50 and 60, this difference will be positive but if the number is between 40 and 50, this difference will be negative. For example, if the number is 57, difference will be 57 – 50 = 7 and if the number is 44, the difference will be 44 – 50 = -6.
    (b) Add the difference found in step 2(a) to 25. This will be the LHS of the answer.
3. For computing the RHS: find the square of the difference found in step 2(a). But remember to give your answer as a two-digit number. For example, if the number is 53, the difference is 3. The square of 3 is 9 but it will be written as 09. This will form the RHS of the answer.
4. Combine the LHS and RHS to get the final answer.

Example 8: Find the square of 57.

Step 1: Find the difference between the number and 50.

57 – 50 = 7.

Step 2: Add the difference to 25 to get the LHS of the answer.

LHS: 25 + 7 = 32.

Step 3: Find the square of the difference. The difference is 7. Square of 7 is 49. This forms the RHS of the answer.

Answer: 57^2 = 3249

Example 9: Find the square of 44.

Step 1: Find the difference between the number and 50.

44 – 50 = -6.

Step 2: Add the difference to 25.

25 + (-6) = 19 ◄------ LHS

Step 3: Find the square of the difference. The difference is (-6). Square of (-6) is 36. This forms the RHS of the answer.

Answer: 44^2 = 1936

Example 10: Find the square of 48.

Step 1: 48 – 50 = -2

Step 2: 25 + (-2) = 23

Step 3: Square of (-2) = 04

Answer: 2304

Example 11: Find the square of 53.

Step 1: 53 – 50 = 3

Step 2: 25 + 3 = 28

Step 3: square of 3 = 09

Answer: 2809

Example 12: Find the square of 56

Answer: 56^2 = 3136

Example 13: Find the square of 42

Answer: 42^2 = 1764

This method can be used for numbers below 40 and above 60 also. The steps remain the same, but the addition and squaring of the difference may be a little more complex, especially when the square of the difference becomes a three-digit number. Let us see it through an example

Example 14: Find the square of 37.

Step 1: 37 – 50 = -13

Step 2: 25 + (-13) = 12

Step 3: Square of (-13) = 169.

Since the RHS must be a two-digit number, we carry forward the 1 from 169 to the LHS.

So, the LHS becomes 13 and RHS is 69.

Answer: $12_1 69$ = 1369.

Example 15: Find the square of 64.

Step 1: 64 – 50 = 14

Step 2: 25 + 14 = 39

Step 3: Square of 14 = 196

Answer: $39_1 96$ = 4096.

10.3 Squares of numbers near the bases.

Steps:

The answer will have two parts – LHS and RHS.

1. For computing the LHS:
   a. Find the difference between the number and the base. In case the number is above the base, the difference will be positive and in case the number is below the base, the difference will be negative. For example, if the number to be squared is 107, the difference will be 107 – 100 = 7 and if we are finding the square of 91, the difference will be 91 – 100 = -9
   b. Add this difference to the number for getting the LHS.
2. For computing the RHS: The RHS will always have the same number of digits as the zeros in the base. For example, if we are squaring 107, the RHS will have two digits as 107 is close to the base 100 and 100 has two zeros. RHS is computed by squaring the difference found in step 1(a).

Example 16: Find the square of 107.

107 is close to 100. Hence the selected base is 100.

Step 1: Find the difference between the number and the base:

107 – 100 = 7.

Step 2: Add this difference to the number:

107 + 7 = 114. This is the LHS of our answer

Step 3: The RHS will have 2 digits. Square the difference. The difference is 7. Square of 7 is 49. This is the RHS of the answer.

Combining the LHS and RHS we get

107^2 = 11449

Example 17: Find the square of 91.

91 is close to 100. Hence the selected base is 100.

Step 1: Find the difference between the number and the base:

91 – 100 = -9.

Step 2: Add this difference to the number:

91 + (-9) = 82. This is the LHS of our answer

Step 3: The RHS will have 2 digits. Square the difference. The difference is 9. Square of 9 is 81. This is the RHS of the answer.

Combining the LHS and RHS we get

91^2 = 8281

Example 18: Find the square of 994.

Since 994 is close to 1000, the selected base will be 1000 and the RHS will have 3 digits.

Step 1: 994 – 1000 = -6

Step 2: 994 + (-6) = 988

Step 3: Square of the difference, $(-6)^2$ = 36. But this will be written as 036.

Answer: 988036

Example 19: Find the square of 1010

Step 1: 1010 – 1000 = 10

Step 2: 1010 + 10 = 1020

Step 3: 10^2 = 100

Answer: 1020100

10.4 Squaring of any number – General method

Till now we have used methods which apply to specific numbers but now we will understand a method which will apply to all numbers. For this the rule used is "dwanda yog" or duplex combination.

Concept of Duplex or D:

1. **Duplex of a single digit number** is square of that number. For a single digit number, a, the Duplex is a^2.

D (a) = a^2,

D (3) = 32 = 9

D (5) = 25 and so on.

2. **Duplex of a two-digit number** is defined as the "twice the product of its digits". For a two-digit number, ab, the D (ab) = 2 x a x b = 2ab

D (53) = 2 x 5 x3 = 30

D (14) = 2 x 1 x 4= 8

3. **Duplex of a three-digit number** is defined as "twice the product of 1st and 3rd digit, plus the square of the second digit. For a three-digit number, abc,

D (abc) = a x c + b^2 = ac + b^2

4. **Duplex of a four-digit number** is defined as twice the product of first and the last digit added to the product of the two middle digits. For a four-digit number, abcd,

D (abcd) = 2 x a x d + 2 x b x c = 2ad + 2bc

D (4326) = 2x 4 x 6 + 2 x 3 x 2 = 48 + 12 = 60

D (1072) = 2 x 1 x 2 + 2 x 0 x 7 = 4 + 0 = 4

5. **Duplex of five-digit number** is twice the product of first and last digit, plus twice the product of second and fourth digit added to the square of the 3rd digit. For a five-digit number, abcde,

D (abcde) = (2 x a x e) + (2 x b x d) + c^2 = 2ae + 2bd + c^2

D (34156) = (2 x 3 x 6) + (2 x 4 x 5) + 1^2 = 36 + 40 + 1 = 77

D (76084) = (2 x 7 x 4) + (2 x 6 x 8) + 0^2 = 56 + 96 = 152

6. **Duplex of a 6-digit number**, abcde, is defined as

D (abcdef) = 2 x a x f + 2 x b x e + 2 x c x d

D (321453) = 2 x 3 x 3 + 2 x 2 x 5 + 2 x 1 x 4 = 18 + 20 + 8 = 46

D (703490) = 2 x 7 x 0 + 2 x 0 x 9 + 2 x 3 x 4 = 24

Summary of Duplex computation

D (a)	a^2
D (ab)	2 x a x b
D (abc)	2 x a x c + b^2
D (abcd)	2 x a x d + 2 x b x c
D (abcde)	2 x a x e + 2 x b x d + c^2
D (abcdef)	2 x a x f + 2 x b x e + 2 x c x d

It is suggested that the students practice the calculation of Duplex of following numbers before proceeding on.

Q. Find the Duplex of:

(a) 254 (b) 371 (c) 2189 (d) 1053 (e) 45231

Answers:

(a) 41 (b) 55 (c) 52 (d) 6 (e) 42

Finding the Square of a given number

Note: If a number consists of n digits, then its square will have 2n or 2n – 1 digits. For example, a two-digit number will have its square as a 4-digit or a 3-digit number. Square of 12 is 144 whereas square of 97 is 9409.

Method to find the square of a n-digit number: Starting from the right, we find the duplex of the digit in unit's place. In the next step we find the duplex of the number formed by digits in ten's place and unit's place. In the next step we find the duplex of the number formed by digits in hundred's place, ten's place and unit's place. Like this at every stage, we keep on including one more digit, till all the digits are involved. After that we start excluding one digit at a time from the right till, we reach the left most digit.

At each stage we will write only 1 digit as the answer and carry forward the rest of the digits to the next step.

Let us understand it for a 2-digit number:

The answer will come in 3 parts – RHS, middle part, and LHS.

If ab is the two-digit number, its square is given by the rule:

LHS = D (a)

Middle part = D (ab)

RHS = D (b)

Combining the three parts, we get our final answer.

Example 20: Find the square of 38

Here a = 3 and b = 8.

Since it is a two-digit number, its square will either have 3 digits or 4 digits.

Steps	**Answer**
Step 1: We start at the right most digit and find its Duplex D. D (8) = 64. From this answer, we write only 4 at the unit's place and carry the 6 forward to the next step.	— — — $_6\underline{4}$
Step 2: We include the next digit 3 and find the Duplex of 38 D (38) = 2 x 3 x 8 = 48. Adding the 6 from previous step we get 54. We again write 4 in the next space of our answer and carry forward 5 to the next step.	— — $_5\underline{4}$ $\underline{4}$
Step 3: Now all the digits in the number have been covered. Now we start excluding one digit at a time from the right.	

Exclude the digit 8 from the right and find the Duplex of 3.

D (3) = 9. Add 5 from the previous step we get 14. That will fill up the next two spaces of our answer.

1 4 4 4

Answer 38^2 = 1444

Example 21: Find the square of 79.

Steps	Answer
Step 1: We start at the right most digit and find its Duplex D. D (9) = 81. From this answer, we write only 1 at the unit's place and carry the 8 forward to the next step.	_ _ _ $_{8}$1
Step 2: We include the next digit 7 and find the Duplex of 79 D (79) = 2 x 7 x 9 = 126. Adding the 8 from previous step we get 134. We again write 4 in the next space of our answer and carry forward 13 to the next step.	_ _ $_{13}$4 1
Step 3: Now all the digits in the number have been covered. Now we start excluding one digit at a time from the right. Exclude the digit 9 from the right and find the Duplex of 7. D (7) = 49. Add 13 from the previous step we get 62. That will fill up the next two spaces of our answer.	6 2 4 1

Answer: 6241

Example 22: Find the square of 83.

Step 1: D (3) = 9

Step 2: D (83) = 2 x 8 x 3 = 48 = $_{4}8$

Step 3: D (8) = 64; 64 + 4 = 68

Answer: 6889

Let us see some examples with 3-digits.

Example 23: Find the square of 321.

Since it is a three-digit number, its square will have 5 or 6 digits.

We will start with the right most digit and find its Duplex and we will keep including one digit at a time till all the three digits are included and then we will exclude one digit at a time. While including, we include digit to the left of the current digit and while excluding, we exclude from the right.

Steps	**Answer**
Step 1: D (1) = 1	_ _ _ _ _ 1
Step 2: D (21) = 2 x 2 x 1= 4	_ _ _ _ 4 1
Step 3: D (321) = 2 x 3 x1 + 2^2 = 10 = $_{1}0$	_ _ _ $_{1}0$ 4 1

At this point we have reached the step where all the digits of the number have been included for the computation of the Duplex. Now we will start excluding the digits from the right side.

Step 4: D (32) = 2 x 3 x 2= 12	
12 + 1 (carry forward) = 13 = $_{1}3$	_ _ $_{1}3$ 0 4 1
Step 5: D (3) = 9	
9 + 1(carry forward) = 10	1 0 3 0 4 1

Answer: 321^2 = 103041.

Example 24: Find the square of 578.

Step 1: D (8) = 64 = $_{6}4$	_ _ _ _ _ $_{6}4$
Step 2: D (78) = 2 x 7 x 8 = 112	
112 + 1 = 113 = $_{11}3$	_ _ _ _ $_{11}3$ 4
Step 3: D (578) = 2 x 5 x 8 + 7^2 = 80 + 49 = 129	
129 + 11 = 140 = $_{14}0$	_ _ _ $_{14}0$ 3 4

Step 4: D (57) = 2 x 5 x 7 = 70

70 + 14 = 84 = $_{8}4$ $\quad$ $\_\,\_\,{}_{8}\underline{4}\,\underline{0}\,\underline{3}\,\underline{4}$

Step 5: D (5) = 25

25 + 8 = 33 $\quad$ $\underline{3}\,\underline{3}\,\underline{4}\,\underline{0}\,\underline{3}\,\underline{4}$

Answer: 578^2 = 334034

Example 25: Find the square of 643

Combining all the steps, we can write it as:

Starting from the right most digit and calculating the Duplex at every stage we get:

643^2 = D (6) / D (64) /D (643) / D (43) / D (3)

= 36 / $_{4}8$ / $_{5}2$ / $_{2}4$ / 9

= 4 1 3 4 4 9

Answer: 643^2= 413449

We can now apply the same method to find the square of a 4-digit number.

Example 26: Find the square of 4136

Since it is a 4-digit number, its square will have 7 or 8 digits.

Step 1: D (6) = 36 = $_{3}6$ $\quad$ $\_\,\_\,\_\,\_\,\_\,\_\,\_\,{}_{3}\underline{6}$

Step 2: D (36) = 2 x 3 x 6 = 36

36 + 3 = 39 = $_{3}9$ $\quad$ $\_\,\_\,\_\,\_\,\_\,\_\,{}_{3}\underline{9}\,\underline{6}$

Step 3: D (136) = 2 x 1 x 6 + 3^2 = 12 + 9 = 21

21 + 3 = 24 = $_{2}4$ $\quad$ $\_\,\_\,\_\,\_\,\_\,{}_{2}\underline{4}\,\underline{9}\,\underline{6}$

Step 4: D (4136) = 2 x 4 x 6 + 2 x 1 x3 = 48 + 6 = 54

54 + 2 = 56 = $_{5}6$ $\quad$ $\_\,\_\,\_\,\_\,{}_{5}\underline{6}\,\underline{4}\,\underline{9}\,\underline{6}$

In this step we have included all the four digits of the number. Form next step we will exclude one digit at a time from the right.

Step 5: D (413) = 2 x 4 x 3 + 1^2 = 25

25 + 5 = 30 = $_{3}0$ $\qquad$ _ _ _ $_{3}$0 6 4 9 6

Step 6: D (41) = 2 x 4 x 1 = 8

8 + 3 = 11 = $_{1}1$ $\qquad$ _ _ $_{1}$1 0 6 4 9 6

Step 7: D (4) = 16

16 + 1 = 17 $\qquad$ 1 7 1 0 6 4 9 6

Answer: 4136^2 = 17106496

Example 27: Find the square of 3764

We can write:

3764^2 = D (3) / D (37) / D (376) / D (3764) / D (764) / D (64) / D (4)

= 9 / $_{4}2$ / $_{8}5$ / $_{10}8$ / $_{9}2$ / $_{4}8$ /$_{1}6$

= 14167696

Answer: 3764^2 = 14167696

This method can be extended to find the square of five-digit or a six-digit number also as shown in the examples below:

Example 28: Find the square of 51237

51237^2 = D (5) / D (51) / D (512) / D (5123) / D (51237) / D (1237) / D (237) / D (37) / D (7)

= 25 / $_{1}0$ / $_{2}1$ / $_{3}4$ / $_{8}0$ / $_{2}6$ / $_{3}7$ / $_{4}2$ / $_{4}9$

= 2625230169

Answer: 51237^2 = 2625230169

Example 29: Find the square of 302161

302161^2 = D (3) / D (30) / D (302) / D (3021) / D (30216) / D (302161) / D (02161) / D (2161) / D (161) / D (61) / D(1)

= $9 / 0 / {}_{1}2 / 6 / {}_{4}0 / {}_{1}0 / {}_{2}5 / {}_{1}6 / {}_{3}8 / {}_{1}2 / 1$

= 91301269921

Summary of the methods for squaring

Case 1: When the number ends in 5

1. The answer comes in two parts – LHS and RHS.
2. For LHS multiply the digit other than 5 by its successor.
3. RHS is always 25.

Case 2: When the numbers are between 40 – 60

1. The answer will have two parts – LHS and RHS
2. For computing the LHS:
   (a) Find the difference between the number and 50. If the number is between 50 and 60, this difference will be positive but if the number if between 40 and 50, this difference will be negative.
   (b) Add the difference found in step 2(a) to 25. This will be the LHS of the answer.
3. For computing the RHS: find the square of the difference found in step 2(a). But remember to give your answer as a two-digit number.
4. Combine the LHS and RHS to get the final answer.

Case 3: When the numbers are near the base.

The answer will have two parts – LHS and RHS.

1. For computing the LHS:
   b. Find the difference between the number and the base. In case the number is above the base, the difference will be positive and in case the number is below the base, the difference will be negative.
   c. Add this difference to the number for getting the LHS.

2. For computing the RHS: The RHS will always have the same number of digits as the zeros in the base. RHS is computed by squaring the difference found in step 1(a).

Case 4: General method for squaring 2-digit, 3-digit, 4-digit numbers

Starting from the right, we find the duplex of the digit in unit's place. In the next step we find the duplex of the number formed by digits in ten's place and unit's place. In the next step we find the duplex of the number formed by digits in hundred's place, ten's place and unit's place. Like this at every stage, we keep on including one more digit, till all the digits are involved. After that we start excluding one digit at a time from the right till, we reach the left most digit.

At each stage we will write only 1 digit as the answer and carry forward the rest of the digits to the next step.

Try it yourself

Q1. Find the squares of the following numbers:

A.	(a) 55	(b) 65	(c) 85	(d) 95	(e) 115
B.	(a) 42	(b) 53	(c) 58	(d) 59	(e)48
	(f) 46	(g) 56	(h) 51	(i) 49	(j) 47
C.	(a) 106	(b) 109	(c) 104	(d) 108	(e) 112
D.	(a) 91	(b) 93	(c) 88	(d) 92	(e) 94
E.	(a) 1004	(b) 1009	(c) 1011	(d) 1003	(e) 1012
F.	(a) 998	(b) 995	(c) 997	(d) 988	(e) 991
G.	(a) 78	(b) 65	(c) 521	(d) 478	(e) 674
	(f) 4111	(g) 3078	(h) 5003	(i) 8091	(j) 9234

Chapter 11

Square roots

Many mathematical operations have an inverse operation. For example, addition has its inverse in subtraction. The inverse of multiplication is division. Squaring of a number, learnt in the previous chapter, has its inverse in square root. The square root of a number, n, written as $\sqrt{n}$ is defined as that number which when multiplied by itself will give n. For example,

$\sqrt{16} = 4$ because 4 x 4 = 16

Although (-4) x (-4) is also 16 but when we write $\sqrt{16}$, we write only the positive square root of the number. Only in case we are solving a quadratic equation in

x, for example, $x^2 = 16$, we write $x = \pm 4$

We write ± 4 as the square of (-4) is also 16. Although the quadratic equation in x will give us two answers, one positive and one negative but for practical purposes, we will consider only the positive answers in this chapter.

In the previous chapter we have seen how easy it is to find the squares of numbers using the techniques of Vedic math. In this chapter we will learn techniques to find square roots of numbers. We will study this in two sections:

1. Square root of perfect squares

2. Square root of imperfect squares.

Perfect Squares

The squares of whole numbers are called perfect squares. For example, 1, 4, 9, 16, 25 are perfect squares as they are obtained on squaring 1, 2, 3, 4 and 5 respectively.

There are rules to identify if a given number is a perfect square.

1. For a number to be a perfect square, its last digit must be 0, 1, 4, 5, 6 or 9. A perfect square ends in 0, 1, 4, 5, 6, 9. But remember not all numbers

ending in these digits are perfect squares. For example, 24 ends in 4 but is not a perfect square.

2. A perfect square never ends with 2, 3, 7 or 8.
3. It should have even number of 0's in the end (in case it ends in a 0).
4. If the number ends in a 6, the digit before 6 must be odd.
5. If the number doesn't end with 6, then the digit before the last digit must be even.
6. If the number is even, then the last two digits must be divisible by 4.

Let us look at a few examples and use the rules to identify if the given number is a perfect square.

Example 1: 144.
The given number ends in 4 and the last two digits are divisible by 4 (Rule 1 and 6). So, it is a perfect square. Its square root is 12.

Example 2: 278
Since the number ends in 8, it cannot be a perfect square (Rule 2).

Example 3: 961
The number ends in 1 and the second last digit is even (Rule 1 and 5). Hence it is a perfect square.

Example 4: 7744
The number ends in 4 and the last two digits are divisible by 4 (Rule 1 and 6). Hence it is a perfect square.

11.1 Square root of perfect squares

To find the square root of perfect squares, we will study the square roots of the first 10 natural numbers and their last digits.

Number	Square	Last digit of the Square
1	1	1
2	4	4
3	9	9
4	16	6
5	25	5
6	36	6
7	49	9
8	64	4
9	81	1
10	100	0

Table 11.1

Observing the above table, we can note the following:

If the last digit of a number is 1, the last digit of its square root could be 1 or 9 (because 1 x 1 = **one** and 9 x 9 = eighty-**one**).

If the last digit of a number is 4, the last digit of its square root could be 2 or 8 (because 2 x 2 = **four** and 8 x 8 = sixty-**four**).

If the last digit of a number is 6, the last digit of its square root could be 4 or 6 (because 4 x 4 = **six**teen and 6 x 6 = thirty-**six**)

If the last digit of a number is 5, the last digit of its square root will be 5 (because 5 x 5 = twenty-**five**)

If the last digit of a number is 0, the last digit of its square root will be 0 (because 10 x 10 is 10**0**)

Summarizing this information, we can make the following table:

Last digit of the perfect square	Last digit of its square root
1	1 or 9
4	2 or 8
9	3 or 7
6	4 or 6
5	5
0	0

Table 11.2

Steps to find the square root of a perfect square:

Before we understand the steps for finding the square root of a perfect square, I have given a table below which lists the squares of multiples of 10 up to 100. This table is extremely helpful in finding the square root.

Number	Square
10	100
20	400
30	900
40	1600
50	2500
60	3600
70	4900
80	6400
90	8100
100	10000

Table 11.3

Step 1: Determining the number of digits in the square root. The number of digits in the square root will be equal to the number of groups formed in Step 1. We can also say that if the number has n digits then its square root will have $\frac{n}{2}$ digits if n is even and $\frac{n+1}{2}$ digits if n is odd. For example, if a number has 7 digits, its square root will have $\frac{7+1}{2} = 4$ digits.

Step 2: Determine the last digit of the square root: Using the table 11.2, we can determine the last digit of the square root. For example, if we are finding the square root of 4624, we know that the last digit of the square root will be either 2 or 8 and the answer at this stage will be _2 or _8.

Step 3: From the table 11.3, determine the squares between which this number lies. For example, if we are finding the square root of 4624, from table 11.3, we can see that 4624 lies between 3600 and 4900. Since the square root of 3600 is 60 and the square root of 4900 is 70, the square root of 4624 will lie between 60 and 70.

Step 4: Combining the steps 2 and 3, we can say that the square root of 4624 will be either 62 or 68.

Step 5: Find the square of the number ending in 5 between the two possible answers determined in step 4. For example, in this case, the two possible answers are 62 and 68, the number ending in 5 between them will be 65. Using the technique learnt in the previous chapter (Section 10.1), we find its square which will be $65^2 = 4225$. Since the given number 4624 > 4225, we will choose the higher of the two options in step 4. Hence the answer is 68 for this example.

Let us understand these steps with the help of more examples.

Example 5: Find the square root of 2116

Step 1: The number of digits in the square root will be 2.

Step 2: Since the number ends in 6, its square root will end in either 4 or 6.

Step 3: From table 11.3, we can see that the given number lies between 1600 and 2500. Since the square root of 1600 is 40 and the square root of 2500 is 50, the square root of 2116 will lie between 40 and 50.

Step 4: Combining step 2 and 3, we can say that the square root of 226 is either 44 or 46.

Step 5: A number ending in 5 between the two possible answers is 45. The square of 45 is 2025. Since 2116 > 2015, we choose the higher option.

Answer: $\sqrt{2116} = 46$

Example 6: Find the square root of 2809.

Step 1: The number of digits in the square root will be 2.

Step 2: From tables 11.2 and 11.3, we can conclude that the answer will be either 53 or 57.

Step 3: The square of number ending in 5 between 53 and 57 is $55^2 = 3025$.

Step 4: Since the given number, 2809 < 3025, we choose the lower option. Hence the answer is 53.

Answer: $\sqrt{2809} = 53$

NOTE: If the given number is a five digit or a six-digit number, the table 11.3 can be extended to figure out the square root or we can also use the general method for finding the square root described in the next section.

11.2 Square root of Imperfect squares.

The general method of finding the square root of a number is very similar to the long division learnt in chapter 9, but with a slight variation.

In this case the divisor is the double of the first digit of the square root and at every stage we will be subtracting the Duplex of the digits after the first digit of the square root from the new dividend.

These steps may not be very clear at the moment but let us understand the steps with the help of an example.

Example 7: Find the square root of 58345.

This number has 5 digits, so its square root will have $\frac{5+1}{2} = 3$ digits.

Step 1: Starting from the right most digit, make groups of 2 digits and if a single digit remains on the left, like in this case, that will be taken as a group also. Hence, in this case, we get 3 groups, namely, 45, 83 and 5.

Step 2: 5 is the first group from the left. Draw a vertical line after 5 as shown.

5	83 45

Step 3: Looking at the first group from the left, 5, we must find the highest perfect square lower than it. In this case, 4 is the highest perfect square which is less than 5. Square root of 4 is 2. So, we write this 2 as the first digit of the required square root. Difference of 5 and the highest perfect square less than it is 1. 5 – 4 = 1. This difference is written with the next group as shown and 18 is our **gross dividend (GD)**.

Double the first digit of the square root to get 4. Write this 4 as the divisor. This is the first variation from the long division. **The divisor is the double of the first digit of the square root.**

4	5	$_1$83	45
	2		

Step 4: Divide the gross dividend, 18 by 4 to get 4 as quotient and 2 as remainder. Write this quotient 4 as the second digit of the required square root. Carry the remainder to the next digit 3 and make it 23 as the next GD.

4	5	$_{1}8$ $_{2}3$	4 5
	2	4	

Step 5: From GD, 23, we will subtract the duplex of the digit to the right of the vertical line. In this case it is 4. Duplex of 4 = 4^2 = 16. The number obtained after subtraction will be our **net dividend (ND). Note that when we divide for the first time, the Gross dividend and the net dividend are same.**

23 – 16 = 7. 7 is the **net dividend**.

Step 6: Divide the net dividend, 7, by the divisor, 4 to get quotient as 1 and remainder as 3.

This quotient becomes the next digit of our required square root. At this point we have the three digits of the square root as 241. Hence, we will put a decimal point after 1. The remainder 3 will be carried on to the next digit to get the GD 34

4	5	$_{1}8$ $_{2}3$	$_{3}4$ 5
	2	4 1•	

Step 7: From the GD, 34, subtract the duplex of the digits to the right of the vertical column.

Duplex of 41 = 2 x 4 x 1 = 8.

34 – 8 = 26. 26 is our ND.

Step 8: Divide the ND, 26, by the divisor 4 to get quotient as 6 and remainder as 2. Write the quotient as he first digit after the decimal point and carry the remainder to the next digit to make it 25 as the next GD.

4	5	$_{1}8\ _{2}3 \quad _{3}4\ _{2}5$
	2	4 1• 6

Step 9: From the GD, 25, subtract the duplex of the digits to the right of the vertical line.

D (416) = 2 x 4 x 6 + 1^2 = 48 + 1 = 49

25 – 49 = -24. Since the dividend cannot be negative, we will reduce the last digit in the answer by 1, making the answer as 241.5, just like we did in case of long division.

So, now the remainder to be carried to the next digit will be 2 + 4 = 6.

Subtracting the duplex of 415 from 65, we get

65 – D (415) = 65 – [2 x 4 x 5 + 12] = 65 – 41 = 24.

24 is our ND.

4	5	$_{1}8\ _{2}3 \quad _{3}4\ _{6}5$
	2	4 1• 5

Step 10: Dividing ND, 24, by the divisor 4 we get 6 as quotient and 0 as remainder. The quotient digit is written as the next digit of the answer.

So, the answer is 241.56.

$$\sqrt{58345} = 241.56$$

For taking out the square root of an imperfect square we put a vertical line after the first group from the left. At each step, we subtract the duplex of all the digits after the vertical line from the corresponding dividend to get the net dividend. The net dividend is then divided by the divisor. And the remainder carried forward to the next digit to get the next GD. This process of division and subtraction continues.

Let us see one more example.

Example 8: Find the square root of 374988.

The given number has 6 digits. The square root will, therefore, have 3 digits followed by a decimal point.

Step 1: Starting from the right, we make groups of two digits each.

Step 2: The first group from the left will be 37. Draw a vertical line after 37.

37	49 88

Step 3: We must find the highest perfect square less than the first group from the left. The highest perfect square less than 37 is 36. And the square root of 36 is 6. So, 6 is our first digit of the required square root. The difference of 37 and the highest perfect square less than 37 is 1. This 1 is written with the next digit to get our GD. Since there is no digit after the vertical line, this is also our ND, which is 14 in this case.

The divisor will be double of the first digit of the square root which means that in this case the divisor will be 12.

12	37	$_{1}49$ 88
	6	

Step 4: Divide ND by the divisor 12 to get quotient as 1 and remainder as 2. The quotient goes as the next digit of the square root and the remainder is written with the next digit of the number to get the GD as 29.

12	3 7	$_{1}4\ _{2}9\quad 8\,8$
	6	1

Step 5: From the GD, 29, subtract the duplex of the digit to the right of the vertical line to get the ND.

29 – D (1) = 29 – 1= 28. 28 is the ND.

Step 6: Divide ND by the divisor, 12, to get the quotient as 2 and the remainder as 4. Quotient 2 is the next digit in our answer and 4 is written with the next digit of the number to get the GD as 48.

Note that this 2 will complete the 3 digits of the square root. Hence, we will put a decimal point after 2.

12	3 7	$_{1}4\ _{2}9\quad _{4}8\,8$
	6	1 2•

Step 7: From the GD, subtract the duplex of the digits to the right of the vertical line to get the ND.

48 – D (12) = 48 – (2 x 1 x 2) = 48 – 4 = 44.

44 is the ND.

Step 8: Divide the ND by the divisor to get quotient 3 and remainder 8. The quotient forms the next digit of the answer and the remainder is written with the next digit of the number to get the GD as 88.

12	3 7	$_{1}4\ _{2}9\quad _{4}8\ _{8}8$
	6	1 2• 3

Step 9: From the GD subtract the duplex of the digits next to the vertical line to get the ND.

88 – D (123) = 88 – [2x1x3 + 22] = 88 – 10 = 78.

78 is the new ND.

Step 10: Divide ND by divisor to get the quotient as 6 and the remainder as 6.

This quotient forms the next digit of our answer.

Answer: $\sqrt{374988} = 612.36$

NOTE: In case the question requires you to find the square root up to three or four decimal places, you may put zeroes at the end of the number and continue the same process to get further digits.

Example 9: Find the square root of 302549

10	30	$_5 2 5$	49
	5		

The first group on the left is 30. The highest perfect square less than 30 is 25. Square root of 25 is 5. 5 is the first digit of the answer and double of 5 is our divisor. 30 – 25 = 5, which is written next to 2 to give 52 as GD. (First step GD = ND).

10	30	$_5 2 5$	49
	5	5	

ND ÷ divisor; 52 ÷ 10 gives 5 as quotient (next digit of answer) and 2 as remainder (goes to the next column to give 25 as GD

10	30	$_5 2\, _2 5$	49
	5	5 0•	

GD – D (5) = 25 – 5^2 = 25 – 25 = 0

ND is 0. ND ÷ divisor gives 0 as quotient and 0 as remainder. GD is now 04. Decimal after 0

10	30	$_5 2\, _2 5$	$_0 4\, _4 9$
	5	5 0•	0

GD – D (50) = 04 – 2 x 5 x 0 = 4

ND is 4. ND ÷ divisor gives 0 as quotient and 4 as remainder. GD is now 49.

10	30	$_5 2\, _2 5$	$_0 4\, _4 9$
	5	5 0•	0 4

GD – D (500) = 49 – (2 x 5 x 0 + 0^2) = 49

ND is 49. ND ÷ divisor gives 4 as quotient and 9 as remainder.

Answer: $\sqrt{302549} = 550.04$

Example 10: Find the square root of 428396

12	42	$_6$8 3 9 6
	6	

The first group on the left is 42. The highest perfect square less than 42 is 36. Square root of 36 is 6. 6 is the first digit of the answer and double of 6 is our divisor. 42 – 36 = 6, which is written next to 8 to give 68 as GD. (First step GD = ND).

12	42	$_6$8 $_8$3 9 6
	6	5

ND ÷ divisor; 68 ÷ 12 gives 5 as quotient (next digit of answer) and 8 as remainder (goes to the next column to give 23 as GD

12	42	$_6$8 $_8$3 $_{10}$9 6
	6	5 4•

GD – D (5) = 83 – 5^2 = 83 – 25 = 58

ND is 58. ND ÷ divisor gives 4 as quotient and 10 as remainder. GD is now 109. Decimal after 4

12	42	$_6$8 $_8$3 $_{10}$9 $_9$6
	6	5 4• 5

GD – D (50) = 109 – 2 x 5 x 4 = 109 – 40 = 69

ND is 69. ND ÷ divisor gives 5 as quotient and 9 as remainder. GD is now 96.

12	42	$_6$8 $_8$3 $_{10}$9 $_9$6
	6	5 4• 5 2

GD – D (545) = 96 – (2 x 5 x 5 + 4^2) = 96 – 66 = 30

ND is 30. ND ÷ divisor gives 2 as quotient and 6 as remainder.

Answer: $\sqrt{428396} = 654.52$

11.3 Square root of decimal numbers

In a decimal number, the digits to the left of the decimal are called integral part of the number. When there is a decimal number, only the digits to the left of the decimal, or the integral part, are considered for number of digits in the square root. For example, 8342.578 will have only 2 digits in the square root as the number of digits in the integral part are 4. Hence the decimal in the answer will come after 2 digits. We make groups of the integral part only for calculating the answer.

18	83	$_{2}4\,2$	.578
	9		

The first group on the left is 83. The highest perfect square less than 83 is 81. Square root of 81 is 9. 9 is the first digit of the answer and double of 9 is our divisor. 83 – 81 = 2, which is written next to 4 to give 24 as GD. (First step GD = ND).

18	83	$_{2}4\,_{6}2$	.578
	9	1•	

ND ÷ divisor; 24 ÷ 18 gives 1 as quotient (next digit of answer) and 6 as remainder (goes to the next column to give 62 as GD. Decimal after 1 in the answer.

18	83	$_{2}4\,_{6}2$	.$_{7}$578
	9	1•3	

GD – D (1) = 62 – 1^2 = 61

ND is 61. ND ÷ divisor gives 3 as quotient and 7 as remainder. GD is now 75.

18	83	$_{2}4\,_{6}2$	.$_{7}5_{15}78$
	9	1•33	

GD – D (13) = 75 – 2x 1x 3 = 69

ND is 69. ND ÷ divisor gives 3 as quotient and 15 as remainder. GD is now 157.

18	83	$_{2}4\,_{6}2$	.$_{7}5_{15}78$
	9	1•337	

GD – D (13) = 157 – (2 x 1 x 3 + 3^2) = 157 – 15 = 142

ND is 141. ND ÷ divisor gives 7 as quotient and 16 as remainder. GD is now 168.

Answer up to three places of decimal is 91.337

$$\sqrt{8342.578} = 91.337$$

NOTE: If the number has only decimal part and the integral part is 0, then we start our answer with a decimal point and make groups starting from the right of the decimal. All the steps remain same.

Summary

Step 1: Make pairs of 2 digits each, starting from the right side. If the number of digits is odd, then the single number remaining on the left side will be counted as a group by itself.

Step 2: If the number has n digits, then the number of digits in the square root before the decimal will be given by $\frac{n}{2}$, if n is even and $\frac{n+1}{2}$, if n is odd.

Step 3: Place a vertical line after the first group from the left.

Step 4: Find the highest perfect square less than the first group on the left. The square root of that perfect square forms the first digit of the answer. Double of this digit will be our divisor.

Step 4: Subtract the square of the first digit from the first group and place the remainder with the next digit in the number. This is our Gross Dividend (GD).

Step 5: Now we do alternate division and subtraction steps. Divide the GD by the divisor and write the quotient as the next digit of the answer and place the remainder before the next digit of the number to get the next GD

Step 6: From this GD subtract the Duplex of the second answer digit to get the Net dividend (ND).

Step 7: Now divide this ND by the divisor and write the quotient as the next digit of the answer and place the remainder before the next digit of the number to get the next GD.

Step 8: At each step, the Duplex of the number starting from the second digit of the answer is subtracted from the GD to get the ND. This ND is divided by the divisor and the quotient of this division is written as the next digit of the answer and the remainder is placed before the next digit of the number to get the next GD.

This process continues till all the digits in the number are exhausted.

Points to remember:

1. At any step if we get a negative ND, then we go to the previous step and reduce the quotient by 1 to increase the remainder and hence increase the next GD.
2. Place the decimal point after the required number of digits as calculated in step 2

Try it yourself

Q1. Find the square root of the following perfect squares

a) 9801
b) 5625
c) 1936
d) 3481
e) 1369
f) 3364
g) 6241
h) 7056
i) 3136
j) 8464

Q2. Find the square root of the following numbers

a) 45634
b) 78654
c) 82345
d) 98743
e) 32456
f) 178655
g) 567349
h) 347654
i) 809761
j) 456321

Chapter 12

Cubes and Cube roots

12.1 Cube of a number

When a number is multiplied by itself three times, we get its cube. For example, if we multiply 2 by itself three times, that is, 2 x 2 x 2, we get 8. 8 is called the cube of 2 and mathematically it is written as

$$2^3 = 8$$

Similarly, $5^3 = 125$ and $9^3 = 729$.

12.2 Cube root of a number

Just as square root is the inverse mathematical operation of squaring a number, similarly, cube root is an inverse operation of cube. Finding the cube root of a number is to find the number which has been multiplied by itself three times. Thus if 8 is the cube of 2 then 2 is the cube root of 8. The symbol of cube root of a number *n* is $\sqrt[3]{n}$. Mathematically, we can write:

$$\sqrt[3]{8} = 2$$

Similarly,

$$\sqrt[3]{125} = 5$$

And

$$\sqrt[3]{729} = 9$$

In this chapter we will be learning how to find cube root of a number. Thus, if you are given the number 8, you have to arrive at 2. Vedic mathematics gives a very simple method of finding cube root of perfect cubes. However, please note that this method cannot be used to find cube roots of imperfect cubes.

12.3 Finding Cube root of a perfect cube.

Given in the table below are the numbers from 1 to 10 and their cubes. This table will be used to find cube roots of higher numbers. It is important that this table is learnt thoroughly before proceeding further.

Number	Cube
1	1
2	8
3	27
4	64
5	125
6	216
7	343
8	512
9	729
10	1000

Table 12.1

Observe the underlined digits in table 12.1. Let us try to understand the relationship between the underlined digits in the two columns and see how this information can be used for finding cube roots.

In the first row, the underlined digits are 1 and 1. This implies that if the last digit of a number 1 then the last digit of its cube root will also be 1. In the second row, the underlined numbers are 2 and 8. This means that if the last digit of a number is 8, the last digit of its cube root is 2.

Observing all the rows, we can summarize the information in the following table:

Last digit of the number	Last digit of the cube root
1	1
2	8
3	7
4	4
5	5
6	6
7	3
8	2
9	9
0	0

Table 12.2

From the above table, it can be concluded that all cube roots end with the same digit as their corresponding cubes except for the pairs of 3 & 7 and 2 & 8, which end with each other.

Steps:

Step 1: Split the given perfect cube in two parts in such a way that the RHS has three digits. The LHS will have the remaining digits of the number. For example, if the given number is 79507, it will be split as 79 | 507. The RHS will have 507 and the LHS will have 79.

Step 2: The answer will also be in two parts – LHS and RHS. We can find them in any order. We can find LHS first or RHS first, it doesn't matter.

Step 3: RHS: For computing the RHS of the answer, we look at the last digit of the number and from table 12.2, we find the last digit of the cube root. That will form the RHS of our answer.

Step 4: **LHS**: For computing the LHS, we refer to table 12.1 and see where the LHS of our given number lies in that table. Consider the perfect cube less than the LHS of the given number and its cube root will give us the LHS of our answer.

Let us understand this with the help of an example.

Example 1: Find the cube root of 79507.

Step 1: 7 9 | 507	Splitting the given number such that RHS has 3 numbers.
Step 2: Answer: __ / __	The answer will have two parts
Step 3: __ / <u>3</u>	RHS: The last digit of the given number is 7. From table 12.2, we can see that the last digit of the answer will be 3.

Step 4:

$\underline{4}/\underline{3}$

LHS: The LHS of the given number is 79. From table 12.1 we can see that 79 lies between 64 and 125. The lower of the two numbers is 64. The cube root of 64 is 4. This forms the LHS of our answer.

Answer: $\sqrt[3]{79507} = 43$

Example 2: Find the cube root of 636056.

Step 1:

636 | 056

Splitting the given number such that RHS has 3 numbers.

Step 2:

Answer: __ / __

The answer will have two parts

Step 3:

__ / $\underline{6}$

RHS: The last digit of the given number is 6. From table 12.2, we can see that the last digit of the answer will be 6.

Step 4:

$\underline{8}/\underline{6}$

LHS: The LHS of the given number is 636. From table 12.1 we can see that 636 lies between 512 and 729. The lower of the two numbers is 512. The cube root of 512 is 8. This forms the LHS of our answer.

Answer: $\sqrt[3]{636056} = 86$

Example 3: Find the cube root of 474552

Step 1:

474 | 552

Splitting the given number such that RHS has 3 numbers.

Step 2:

Answer: __ / __

The answer will have two parts

Step 3:

__ / 8

RHS: The last digit of the given number is 2. From table 12.2, we can see that the last digit of the answer will be 8.

Step 4:

7 / 8

LHS: The LHS of the given number is 474. From table 12.1 we can see that 474 lies between 343 and 512. The lower of the two numbers is 343. The cube root of 343 is 7. The LHS of our answer is 7.

Answer: $\sqrt[3]{474552} = 78$

Example 4: Find the cube root of 328509.

328 | 509

↙ ↘

6 9

Answer: $\sqrt[3]{328509} = 69$

Example 5: Find the cube root of 50653

5 0 | 653

↙ ↘

3 7

Answer: $\sqrt[3]{50653} = 37$

Example 6: Find the cube root of 912673

912 | 673

↙ ↘

9 7

Answer: $\sqrt[3]{912673} = 97$

As we can see from the above examples, there are no calculations involved in calculating the cube root of a perfect cube.

Summary of Steps:

1. Split the number into two parts such that the RHS has 3 digits.

2. The answer will have two parts – LHS and RHS. For computing the RHS, we refer to the table 12.2. The last digit of the number will tell us the last digit of the cube root. This digit will be the RHS of our required answer. For computing the LHS, we refer to table 12.1. The LHS of the given number will lie between two perfect cubes. The cube root of the smaller perfect cube will give us the LHS of our answer.

3. The order for finding the LHS and RHS does not matter. We can find either of the digits first.

12.4. Finding the cube of a number

Method I: Formula method

The formula method is the simplest and straight forward method of finding cube of a number. It is particularly useful to find cubes of numbers which are closer to a base.

The two formulae that we will be using are:

$$(a+b)^3 = a^3 + 3a^2b + 3ab^2 + b^3$$

$$(a-b)^3 = a^3 - 3a^2b + 3ab^2 - b^3$$

We can express the given number as a sum or difference of two numbers and then use one of the above formulae to find the cube of that number.

Example 7: Find the cube of 103.

103 can be written as 100 + 3. Hence, substituting a as 100 and b as 3 in the formula 1, we get,

$$(100+3)^3 = 100^3 + 3 \times (100)^2 \times 3 + 3 \times 100 \times 3^2 + 3^3$$

$$= 1000000 + 90000 + 2700 + 27$$

$$= 1092727$$

Example 8: Find the cube of 98

98 can be expressed as (100 – 2). Substituting a as 100 and b as 2 in the second formula, we get,

$$(100-2)^3 = 100^3 - 3 \times (100)^2 \times 2 + 3 \times 100 \times (2)^2 - 2^3$$

$$= 1000000 - 60000 + 1200 - 8$$

$$= 941192$$

Method 2: The *Anurupya Sutra*

The second method that we will learn is *Anurupya Sutra* which is based on the two formulae learnt in the previous section.

Let us look at the first formula

$$(a+b)^3 = a^3 + 3a^2b + 3ab^2 + b^3$$

The expression on the RHS can be broken into two parts as given below. The terms $3a^2b$ has been split into a^2b and $2a^2b$. Similarly, the term $3ab^2$ has been split into ab^2 and $2ab^2$. The whole expression has been written in two lines. The first line has a^3, a^2b, ab^2, b^3. The second line has $2a^2b$ and $2ab^2$.

$(a+b)^3 = a^3 + 3a^2b + 3ab^2 + b^3$

$= a^3 + a^2b + ab^2 + b^3$

$\qquad + 2a^2b + 2ab^2$

Observe the RHS of the first row. The four terms are a^3, a^2b, ab^2 and b^3. If we start at a^3 and multiply it by $\frac{b}{a}$, we will get the second term, a^2b. Similarly, if we multiply the second term by $\frac{b}{a}$, we will get the third term ab^2. And if the third term is multiplied by $\frac{b}{a}$, we will get the last term b^3.

$$a^3 \times \frac{b}{a} = a^2b$$

$$a^2b \times \frac{b}{a} = ab^2$$

$$ab^2 \times \frac{b}{a} = b^3$$

Fig. 12.1

Based on the above three equations, we can infer that going from left to right, the terms in the first row have a ratio of $\frac{b}{a}$ between them. Similarly, going from right to left, the terms will have a reverse ratio of $\frac{a}{b}$ between them.

Also, the terms in the second row of the formula can be obtained by multiplying the terms above them by 2.

Let us now use this formula to find the cube of a number.

$(a+b)^3 = a^3 + a^2b + ab^2 + b^3$ ⟵ First row

$+ 2a^2b + 2ab^2$ ⟵ Second row

Consider a two-digit number 'ab'.

Steps to find its cube:

1. Identify a and b. Compute the ratio $\frac{b}{a}$.
2. Compute the four terms of the first row by using the equations given in figure 12.1 Write the four terms in four columns. The terms computed will be a^3, a^2b, ab^2, b^3.
3. Double the second term a^2b and the third term ab^2 and write the answer in the second row just below the respective terms.
4. We start adding from the right most column and keep writing one digit below each column and carry forward the remaining digits to the next column. The process continues till we have added the left most column.

Example 9: Find the cube of 42.

Step 1: Here a = 4 and b = 2.

The ratio $\frac{b}{a} = \frac{2}{4} = \frac{1}{2}$

Step 2: Finding the four terms of the first row and writing the answer in four columns in a row

$a^3 = 4^3 = \mathbf{64}$

$a^3 \times \frac{b}{a} = 64 \times \frac{1}{2} = \mathbf{32}$

$a^2b \times \frac{b}{a} = 32 \times \frac{1}{2} = \mathbf{16}$

$ab^2 \times \frac{b}{a} = 16 \times \frac{1}{2} = \mathbf{8}$

Writing the answer in a single row in four columns we get

64	32	16	8

Step 3: Computing the second row:

$a^2b = 32$

$2a^2b = 32 \times 2 = \mathbf{64}$

$ab^2 = 16$

$2ab^2 = \mathbf{32}$

Write these in the second row but in the second and third column respectively.

64	32	16	8
	64	32	

Step 4: Starting from the right most column, we add the numbers in each column. We will write only one digit in the answer column and carry forward the other digits to the next column. The first column on the right has only 8. So that will be written in the answer row as it is. In the next column, we have 16 + 32 which gives 48. We write 8 in the answer row and carry forward 4 to the next column. We continue this process till we reach the left most column.

64	32	16	8
	64	32	
74	100	48	8

Answer: $(42)^3$ = 74088 ← Answer row

Example 10: Find the cube of 68

Step 1: $a = 6; b = 8; \frac{b}{a} = \frac{8}{6} = \frac{4}{3}$

Step 2: Computing the first row:

$a^3 = 6^3 = \mathbf{216}$

$a^3 \times \frac{b}{a} = 216 \times \frac{4}{3} = \mathbf{288}$

$a^2b \times \frac{b}{a} = 288 \times \frac{4}{3} = \mathbf{384}$

$ab^2 \times \frac{b}{a} = 384 \times \frac{4}{3} = \mathbf{512}$

216	288	384	512

Step 3: Computing the second row:

$288 \times 2 = 576$

$384 \times 2 = 768$

216	288	384	512
	576	768	

Step 4: Addition

216	288	384	512
	576	768	
314	$_{98}4$	$_{120}3$	$_{51}2$

Answer: $(68)^3$ = 314432

Example 11: Find the cube of 37

Step 1: $a = 3; b = 7; \frac{b}{a} = \frac{7}{3}$

Step 2: $a^3 = 3^3 = \mathbf{27}$

$a^3 \times \frac{b}{a} = 27 \times \frac{7}{3} = \mathbf{63}$

$a^2b \times \frac{b}{a} = 63 \times \frac{7}{3} = \mathbf{147}$

$ab^2 \times \frac{b}{a} = 147 \times \frac{7}{3} = \mathbf{343}$

27	63	147	343

Step 3:

63 x 2 = 126

147 x 2 = 294

27	63	147	343
	126	294	

Step 4:

27	63	147	343
	126	294	
50	$_{23}6$	$_{47}5$	$_{34}3$

Answer: $(37)^3$ = 50653

Some more examples:

Example 12: Find the cube of 12

Here $a = 1; b = 2; \frac{b}{a} = \frac{2}{1} = 2$

a^3	a^2b	ab^2	b^3
1	2	4	8
	4	8	
1	7	$_{1}2$	8

Answer: $(12)^3$ = 1728

Example 13: Find the cube of 51

Here $a = 5; b = 1; \frac{b}{a} = \frac{1}{5}$

a^3	a^2b	ab^2	b^3
125	25	5	1
	50	10	
132	$_{7}5$	$_{1}5$	1

Answer: $(51)^3$ = 132551

Method 3: For numbers near the bases

The method for finding the cubes of numbers near the bases is very similar to the method used for finding the square of a number in chapter 10.

The answer will come in 3 parts – LHS, middle part and the RHS. The LHS can have as many digits but the number of digits in the middle part and the RHS will be same as the number of zeroes in the selected base. If at any stage, the number of digits in the middle part or the RHS exceed the number of zeroes in the base then we carry forward the extra digits to the part of the answer on the left.

Step 1: Finding the LHS: Find the difference between the number and the base. If the number is above the base, the difference will be positive. But if the number is below the base, the difference will come as negative. Note that we need to be careful with the sign of the difference. Add twice of this difference to the number and that will form our LHS.

Step 2: Finding the middle part: Multiply the square of the difference with 3 to get the middle part.

Step 3: Finding the RHS: Find the cube of the difference. That will form the RHS of the answer. In case of numbers below the base, the RHS will be negative and there will be one more step to make it positive.

Example 14: Find the cube of 104.

Selected base will be 100 as 104 is close to 100. So, the RHS and the middle part will have 2 digits each as 100 has 2 zeroes.

Step 1: LHS: Difference between the number and the base:

_112_ /	__ /	__
LHS	middle	RHS

104 – 100 = 4

Number + twice the difference = 104 + (2x4) = 112

Step 2: Middle part: Square of the difference = $(4)^2$ = 16

_112_ /	_48_ /	__
LHS	middle	RHS

3 x 16 = 48. This is the middle part.

Step 3: RHS: cube of the difference = $(4)^3 = 64$

This is the RHS of the answer

Answer: $(104)^3 = 1124864$

Example 14: Find the cube of 98.

Selected base will be 100 as 98 is close to 100. So, the RHS and the middle part will have 2 digits each as 100 has 2 zeroes.

Step 1: LHS: Difference between the number and the base:

98 – 100 = -2

_94_/ __ /__		
LHS	middle	RHS

Number + twice the difference = 98 + [2 x (-2)] = 94

Step 2: Middle part: Square of the difference = $(-2)^2 =$ 4

3 x 4 = 12. This is the middle part.

_94_/ _12_ /__		
LHS	middle	RHS

Step 3: RHS: cube of the difference = $(-2)^3 = -8$

Since this is negative, we will reduce the middle part by 1 and write the complement of the RHS answer. Note that the complement will be calculated with respect to the selected base. Hence the middle part will be 11 and the RHS will be 92.

_94_/ _11_ /_92_		
LHS	middle	RHS

Answer: $(98)^3 = 941192$

Example 15: Find the cube of 106

Selected base will be 100 as 106 is close to 100. So, the RHS and the middle part will have 2 digits each as 100 has 2 zeroes.

Step 1: LHS: Difference between the number and the base:

106 – 100 = 6

_118_/ __ /__		
LHS	middle	RHS

Step 2: Middle part: Square of the difference = $(6)^2 = 36$

_118_/ _$_1$08_ /__

LHS middle RHS

3 x 36 = 108. This is the middle part. But we will write only 08 in the middle part and carry forward 1 to the LHS.

Step 3: RHS: cube of the difference = $(6)^3 = 216$.

_118_/ _$_1$08_ /_$_2$16_

LHS middle RHS

Since the RHS can have only 2 digits, we will write only 16 and carry forward 2 to the middle part.

So, the final answer after adjusting the carry forwards will be:

Answer: $(106)^3$ = 1191016

Example 16: Find the cube of 92.

Selected base will be 100 as 92 is close to 100. So, the RHS and the middle part will have 2 digits each as 100 has 2 zeroes.

Step 1: LHS: Difference between the number and the base:

_76_/ __ /__

LHS middle RHS

92 – 100 = -8

Number + twice the difference = 92 + [2 x (-8)] = 76

Step 2: Middle part: Square of the difference = $(-8)^2 = 64$

_76_/ _$_1$92_ /__

LHS middle RHS

3 x 64 = 192. This is the middle part. We will write 92 and carry 1 to the LHS.

Step 3: RHS: cube of the difference = $(-8)^3 = -512$

We will write negative 12 in the RHS and carry over -5 to the middle part. The middle part becomes 87. Then we take the complement of 12 and reduce the middle part further by 1. Hence, we get 778688 as our final answer.

_77_/ _92_ /_(-512)_

_77_/ _92_/ _$_{(-5)}$(-12)

_77_/_87_/ (complement of 12)

_77_/_86_/88

Answer: $(92)^3$ = 778688

Some more examples:

Example 17: Find the cube of 988

Difference: 988 – 1000 = -12

LHS: 988 – (2 x (-12)) = 964

Middle part: 3 x $(-12)^2$ = 432

(middle part and RHS will have 3 digits each)

RHS: (-12)3 = -1728

Complete answer: 964 / 432 / -($_1$728)

964 / 431 / (complement of 728)

964 / 430 / 272

Answer: 964430272

Example 18: Find the cube of 112

Difference: 112 – 100 = 12

LHS: 112 + (2 x 12) = 136

Middle part: 3 x $(12)^2$ = 432

RHS: (12)3 = 1728

Answer: 136 / $_4$32 / $_{17}$28 = 140 / 49/ 28 = 1404928

Summary of steps for calculating cube of a number

Method 1: Using the formula:

In this method the number is expressed as sum or difference of two numbers and the following formulae are applied:

$$(a+b)^3 = a^3 + 3a^2b + 3ab^2 + b^3$$

$$(a-b)^3 = a^3 - 3a^2b + 3ab^2 - b^3$$

Method 2: General method for finding the cube of a number:

Consider a two-digit number 'ab'.

Steps to find its cube:

1. Identify a and b. Compute the ratio $\frac{b}{a}$.

2. Compute the four terms of the first row by using the equations given in figure 12.1 Write the four terms in four columns. The terms computed will be a^3, a^2b, ab^2, b^3.

3. Double the second term a^2b and the third term ab^2 and write the answer in the second row just below the respective terms.

4. We start adding from the right most column and keep writing one digit below each column and carry forward the remaining digits to the next column. The process continues till we have added the left most column.

Method 3: to find the cube of a number near the base

The answer will come in 3 parts – LHS, middle part and the RHS. The LHS can have as many digits but the number of digits in the middle part and the RHS will be same as the number of zeroes in the selected base. If at any stage, the number of digits in the middle part or the RHS exceed the number of zeroes in the base then we carry forward the extra digits to the part of the answer on the left.

Step 1: Finding the LHS: Find the difference between the number and the base. If the number is above the base, the difference will be positive. But if the number is below the base, the difference will come as negative. Note that we need to be careful with the sign of the difference. Add twice of this difference to the number and that will form our LHS.

Step 2: Finding the middle part: Multiply the square of the difference with 3 to get the middle part.

Step 3: Finding the RHS: Find the cube of the difference. That will form the RHS of the answer. In case of numbers below the base, the RHS will be negative and there will be one more step to make it positive.

Try it yourself

Q1. Find the cube root of the following perfect cubes:

(a) 54872 (b) 117649 (c) 389017 (d) 300763

(e) 205379 (f) 681472 (g) 132651 (h) 39304

Q2. Find the cube root of the following numbers:

(a) 421875 (b) 373248 (c) 658503 (d) 551368

(e) 46656 (f) 970299 (g) 328509 (h) 474552

Q3. Find the cube of the following numbers:

(a) 76 (b) 86 (c) 35 (d) 91 (e) 47

(f) 63 (g) 28 (h) 17 (i) 31 (j) 72

Q4. Find the cube of the following numbers:

(a) 95 (b) 107 (c) 89 (d) 87 (e) 109

(f) 111 (g) 1002 (h) 998 (i) 989 (j) 1004

Chapter 13

Fraction to Decimal

The conventional procedure to convert a fraction to a decimal number is the long division process. We have already seen a shorter approach to long division in chapter 9 where we learnt how to get the quotient in a single line. Vedic mathematics also provides an easy and faster method to convert fractions to decimals. The technique that we are going to learn in this chapter will be useful to convert fractions to decimals when:

(a) the numerator is less than the denominator

(b) the denominator is a small 2-digit number

(c) the denominator is 9, 8 or 1.

For other denominators like 2, 3, 4, 5, 6 or 7 we first multiply the numerator and denominator of the fraction by a suitable number so that the denominator ends in 9, 8 or 1.

The sutra used is "*by one more than one before*". In this method the fraction is first converted to a simpler fraction called the "**Auxiliary Fraction**" (AF). Auxiliary fraction will have simpler divisor and division will be done in one line eliminating the need for long division.

We will first consider fractions whose denominators end in 9, 8 or 1 and then the others.

Case 1: When the denominator is 9.

Let us consider a case where the denominator ends in 9, e.g. 19, 29, 129 etc.

Example 1: Convert $\frac{11}{29}$ to its decimal form.

Step 1: Convert the fraction $\frac{11}{29}$ to its Auxiliary fraction (AF) as given below:

Add 1 to the denominator to make it 30 and then divide the numerator and denominator by 10 to get 3 as the denominator.

A.F. of $\frac{11}{29} = \frac{11}{30} = \frac{1.1}{3}$

This new fraction $\frac{1.1}{3}$ is called Auxiliary fraction (A.F.) of $\frac{11}{29}$

As we can see now our denominator is 3 and not 29. Now we will do step by step division by taking divisor as 3.

Step 2: Divide 1.1 by 3. This will be a slightly different kind of division as $\frac{1.1}{3}$ is an auxiliary fraction and not a normal fraction.

Put a decimal point first and divide numerator 11 by 3. We will get

11 ÷ 3 gives 3 as quotient and 2 as remainder. Write the quotient Q = 3 after the decimal point in the answer and place the remainder R = 2 before the quotient as shown. The number formed by remainder and quotient taken together will give us the Gross dividend (GD) for the next step as 23.

$\frac{1.1}{3} = 0.{}_23$

Step 3: Divide the new GD, 23, by 3. That will give 7 as quotient and 2 as remainder. We write 7 as the next digit in our answer and place the 2 next to 7 to get the next GD as 27.

$\frac{1.1}{3} = 0.{}_23{}_27$

Step 4: Divide the new GD, 27 by 3. It gives 9 as quotient and 0 as remainder. Write the quotient, 9, as the next digit in the answer and place the remainder, 0, next to 9 to get the next GD as 09.

$\frac{1.1}{3} = 0.{}_23{}_27{}_09$

Step 5: Divide the new GD, 09, by 3. That will give us 3 as quotient and 0 as remainder. The quotient digit is the next answer digit and the remainder is placed next to the quotient digit to get the new GD.

$\frac{1.1}{3} = 0.{}_23{}_27{}_09{}_03$

Step 6: Divide the new GD, 03 by 3 to get 1 as quotient and 0 as remainder. Again, placing the quotient and remainder the same way as in previous steps, we get

$\frac{1.1}{3} = 0.{}_23{}_27{}_09{}_03{}_01$

Step 7: the new GD is 01. It cannot be divided by 3. Hence, we place a 0 as the next digit in the answer and take 10 as the next GD.

$\frac{1.1}{3} = 0.{}_23{}_27{}_09{}_03{}_010$

This process can be continued up to any number of decimal places.

Finding the answer up to 15 decimal places, we get

$\frac{11}{29} = \frac{1.1}{3}\,(A.F.) = 0.{}_23{}_27{}_09{}_03{}_010{}_13{}_14{}_24{}_08{}_22{}_17{}_25{}_18{}_06{}_02$

Discarding all the remainders, the final answer can be written as:

$$\frac{11}{29} = 0.3793103448275862$$

We can go on and find up to more decimal places, if need be.

As we can see, this is far simpler than the conventional method of dividing 11 by 29. Let us solve another example.

Example 2: Convert $\frac{13}{59}$ to its decimal form.

Step 1: To find the AF of $\frac{13}{59}$, add 1 to the denominator and then divide the fraction by 10 to get

A.F. of $\frac{13}{59} = \frac{13}{60} = \frac{1.3}{6}$

Step 2: Start the answer with a decimal point and divide 13 by 6. That gives us quotient as 2 and remainder as 1. Write the quotient as the next digit and placing the remainder before the quotient to get the next GD.

13 ÷ 6 gives 2 as quotient and 1 as remainder

$\frac{1.3}{6} = 0.{}_{1}2$

Next GD is 12.

Step 3: Divide 12 by 6 to get 2 as quotient and 0 as remainder.

Next digit of the answer is 2 and the next GD is 02.

$\frac{1.3}{6} = 0.{}_{1}2{}_{0}2$

Step 4: 02 is not divisible by 6. Hence, we place a 0 as the next digit of the answer and take 20 as our GD.

$\frac{1.3}{6} = 0.{}_{1}2{}_{0}20$

Step 5: Divide 20 by 6 we get 3 as quotient and 2 as remainder. Our next digit of the answer is the quotient, 3 and placing the remainder before the quotient gives us the next GD which is 23.

$\frac{1.3}{6} = 0.{}_{1}2{}_{0}20{}_{2}3$

Continuing the same process, we get the answer up to 12 places of decimal as:

$\frac{1.3}{6} = 0.{}_{1}2{}_{0}20{}_{2}3{}_{5}3{}_{5}8{}_{4}9{}_{1}8{}_{0}30{}_{0}50$

Discarding all the remainders, we get

$$\frac{1.3}{6} = 0.220338983050$$

Example 3: Convert $\frac{6}{19}$ to its decimal form.

Step 1:

A.F. of $\frac{6}{19} = \frac{6}{20} = \frac{0.6}{2}$

Step 2: Start the answer with a decimal point. Divide 6 by 2 to get 3 a quotient and 0 as remainder. 3 will be the next digit of the answer and 0 will be placed before 2 to get the GD as 03.

$\frac{0.6}{2} = 0._{0}3$

Step 3: Divide 03 by 2 to get 1 as quotient and 1 as remainder. The quotient becomes the next digit of the answer and the placing the remainder before the quotient gives the next GD as 11.

$\frac{0.6}{2} = 0._{0}3_{1}1$

Step 4: Divide 11 by 2 to get 5 as quotient and 1 as remainder. 5 becomes the next digit of the answer and 1 is placed before the quotient to get the next GD as 15.

$\frac{0.6}{2} = 0._{0}3_{1}1_{1}5$

Continuing this process of division, we get

$\frac{0.6}{2} = 0._{0}3_{1}1_{1}5_{1}7_{1}8_{0}9_{1}4_{0}7_{1}3_{1}6_{0}8_{0}4_{0}2$

Discarding the remainders, we get our final answer as:

$$\frac{6}{19} = 0.3157894736842$$

Example 4: Convert $\frac{25}{39}$ to its decimal form.

A.F. of $\frac{25}{39} = \frac{25}{40} = \frac{2.5}{4}$

Using the steps explained above we get:

$\frac{2.5}{4} = 0._{1}6_{0}4_{0}1_{0}0_{2}2_{2}5_{1}6_{0}4_{0}1$

We can see that after the fifth decimal place, the digits in the answer are repeating. Such decimal numbers are called recurring decimals.

We can write them as:

$\frac{25}{39} = 0.\overline{64102}$, the bar on top indicating that the digits are repeated in the same order.

Case 2: When the denominator ends in 8

When the denominator ends in 8, the steps will almost be the same but since the denominator is 8, which is one less than 9, at each step, after division, the quotient obtained will be added to the gross dividend and then divided further by the denominator of the A.F.

Let us understand this with the help of an example.

Example 5: Convert $\frac{5}{28}$ to its decimal form

Step 1: To find the A.F.: Add 2 to the denominator and divide the fraction by 10

A.F. of $\frac{5}{28} = \frac{5}{30} = \frac{0.5}{3}$

Step 2: This step is same as case 1 when the denominator was 9. Start the answer with a decimal point. Divide 5 by 3 to get 1 as quotient and 2 as remainder. Place the quotient as the first digit of the answer and place the remainder before the quotient to get the Gross dividend. But in this case, since the denominator is 8, which is one less than 9, the quotient will be added to the GD to obtain the final GD.

5 ÷ 3 gives 1 as quotient and 2 as remainder.

$\frac{0.5}{3} = 0.{}_{2}1^{+1}$

Hence, our GD for the next stage is 22.

Step 3: Divide GD by 3 to get 7 as quotient and 1 as remainder. Place the quotient as the next digit of the answer and place the remainder before the quotient. Also add the quotient to the number obtained by the combination of remainder and quotient to get the Gross Dividend.

22 ÷ 3 gives 7 as quotient and 1 as remainder.

Placing the remainder before the quotient we get 17. Adding the quotient to this number gives us: 17 + 7 = 24 as our next GD.

$\frac{0.5}{3} = 0.{}_{2}1^{+1}{}_{1}7^{+7}$

Step 4: Dividing the GD, 24, by 3 we get 8 as the quotient and 0 as remainder. Place the quotient as the next digit of the answer and place the remainder before the quotient to obtain 08. Add the quotient to this number to get the GD.

24 ÷ 3 gives 8 as quotient and 0 as remainder.

08 + 8 = 16. This 16 is our new GD.

$\frac{0.5}{3} = 0.{}_{2}1^{+1}{}_{1}7^{+7}{}_{0}8^{+8}$

Continuing the same process, we get

$\frac{0.5}{3} = 0.{}_{2}1^{+1}{}_{1}7^{+7}{}_{0}8^{+8}{}_{1}5^{+5}{}_{2}6^{+6}$

At this stage we see that the GD for the next step is 32. When we divide this GD by 3, we get 10 as quotient and 2 as remainder. The quotient forms the next digit of the answer, but we cannot have 10 as quotient as it has two digits. So, we increase the previous digit by 1 and write 0 as the next digit of the answer.

$\frac{0.5}{3} = 0._{2}1^{+1}{}_{1}7^{+7}{}_{0}8^{+8}{}_{1}5^{+5}{}_{2}7_{2}0^{+10}$

At the next stage the GD is 20 + 10 = 30.

Dividing this by 3 also gives us 10. We follow the steps as we did in the previous step and write:

$\frac{0.5}{3} = 0._{2}1^{+1}{}_{1}7^{+7}{}_{0}8^{+8}{}_{1}5^{+5}{}_{2}7_{2}1_{0}0^{+10}$

0 + 10 = 10 is our next GD and now we can continue the same process.

Answer:

$$\frac{5}{28} = 0.178571$$

Example 6: Convert $\frac{7}{38}$ to its decimal form.

Step 1: Find the A.F. of $\frac{7}{38}$

A.F. of $\frac{7}{38} = \frac{7}{38+2} = \frac{7}{40} = \frac{0.7}{4}$

Step 2: Start with a decimal and divide 7 by 4.

7 ÷ 4 gives 1 as quotient and 3 as remainder. The quotient forms the first digit of the answer and the remainder is placed before the quotient to get 31. To this number, the quotient is added to get the GD.

31 + 1 = 32. This is the GD for the next step.

$\frac{0.7}{4} = 0._{3}1^{+1}$

Step 3: Divide the GD by 4 to get 8 as quotient and 0 as remainder. The quotient forms the next digit of the answer and place the remainder before the quotient to get 08. To this number we add the quotient to get the next GD as 16.

$\frac{0.7}{4} = 0._{3}1^{+1}{}_{0}8^{+8}$

We continue this process to get:

$\frac{0.7}{4} = 0._{3}1^{+1}{}_{0}8^{+8}{}_{0}4^{+4}{}_{0}2^{+2}{}_{0}1^{+1}0_{0}5^{+5}$

Discarding the remainders and the addition numbers we get,

$$\frac{7}{38} = 0.1842105$$

Example 7: Convert $\frac{65}{78}$ to its decimal form

A.F. of $\frac{65}{78} = \frac{65}{80} = \frac{6.5}{8}$

$\frac{6.5}{8} = 0._{1}8^{+8}{}_{2}3^{+3}{}_{2}3^{+3}{}_{2}3^{+3}{}_{2}3^{+3}{}_{2}3^{+3}{}_{2}3^{+3}$

$\frac{65}{78} = 0.8\overline{3}$

Example 8: Convert $\frac{43}{88}$ to its decimal form.

A.F. of $\frac{43}{88} = \frac{43}{90} = \frac{4.3}{9}$

$\frac{4.3}{9} = 0._{7}4^{+4}{}_{6}8^{+8}{}_{4}8^{+8}{}_{2}6^{+6}{}_{5}3^{+3}{}_{2}6^{+6}{}_{5}3^{+3}{}_{2}6^{+6}{}_{5}3^{+3}{}_{2}6^{+6}{}_{5}3^{+3}$

Discarding the remainders and the additional numbers, we get our final answer as

$$\frac{43}{88} = 0.488636363 = 0.488\overline{63}$$

Case 3: When the denominator ends in 1

Example 9: Convert $\frac{10}{21}$ to its decimal form.

Step 1: To find the Auxiliary fraction (A.F.): When the denominator ends in 1 then to find its Auxiliary fraction, we will subtract 1 from the numerator and the denominator and then divide the numerator and denominator by 10.

A.F. of $\frac{10}{21} = \frac{10-1}{21-1} = \frac{9}{20} = \frac{0.9}{2}$

Step 2: We start with a decimal point and divide 9 by 2.

9 ÷ 2 gives 4 as quotient Q1 and 1 as remainder, R1. We write Q1 as the first digit of our answer and place R1 before the quotient.

$\frac{0.9}{2} = 0._{1}4$

But here there is a slight difference. Instead of taking R1Q1 as our gross dividend (GD) for the next step, we will be taking R1(9-Q1). In this case 9 – Q1 = 9 – 4 = 5. Hence our GD for the next step will be 15.

$\frac{0.9}{2} = 0._{1}4^{5}$

Step 3: Divide the GD, 15, by 2 to get 7 as quotient and 1 as remainder. Subtracting this quotient from 9 we get 9 – 7 = 2. Placing the remainder with this difference gives us 12, which forms our next GD.

$\frac{0.9}{2} = 0._{1}4^{5}{}_{1}7^{2}$

Step 4: Dividing the GD, 12, by 2 we get 6 as the quotient and 0 as remainder. Subtracting this quotient from 9 we get 9 – 6 = 3. Placing the remainder with 3 gives us 03 which is our next GD.

$\frac{0.9}{2} = 0._{1}4^{5}{}_{1}7^{2}{}_{0}6^{3}$

Continuing the same process of dividing, subtracting the quotient from 9, combining with the difference, we get our answer as:

$\frac{10}{21} = 0.{}_{1}4^{5}{}_{1}7^{2}{}_{0}6^{3}{}_{1}1^{8}{}_{0}9^{0}{}_{0}0^{9}{}_{1}4^{5}{}_{1}7^{2}{}_{0}6^{3}{}_{1}1^{8}{}_{0}9^{0}{}_{0}0^{9}{}_{1}4^{5}$

Discarding the remainders and the differences, we get,

$$\frac{10}{21} = 0.476190476190 = 0.\overline{476190}$$

Example 10: Convert $\frac{8}{31}$ to its decimal form.

Step 1: Finding the A.F.

A.F. of $\frac{8}{31} = \frac{8-1}{31-1} = \frac{7}{30} = \frac{0.7}{3}$

Step 2: Start with a decimal point.

$7 \div 3$ gives 2 as quotient (Q1) and 1 as remainder(R1).

Subtracting Q1 from 9,

$9 - 2 = 7$.

Next GD = 17

$\frac{0.7}{3} = 0.{}_{1}2^{7}$

Step 3: $17 \div 3$ gives 5 as quotient(Q2) and 2 as remainder.

$9 - Q2 = 9 - 5 = 4$.

Next GD = 24

$\frac{0.7}{3} = 0.{}_{1}2^{7}{}_{2}5^{4}$

Step 4: $24 \div 3$ gives 8 as quotient(Q3) and 0 as remainder.

$9 - Q3 = 9 - 8 = 1$.

Next GD = 01.

$\frac{0.7}{3} = 0.{}_{1}2^{7}{}_{2}5^{4}{}_{0}8^{1}$

Step 5: $01 \div 3$ gives 0 as quotient (Q4) and 1 as remainder.

$9 - Q4 = 9 - 0 = 9$

Next GD = 19.

$\frac{0.7}{3} = 0.{}_{1}2^{7}{}_{2}5^{4}{}_{0}8^{1}{}_{1}0^{9}$

Continuing the same process, we get,

$\frac{0.7}{3} = 0.{}_{1}2^{7}{}_{2}5^{4}{}_{0}8^{1}{}_{1}0^{9}{}_{1}6^{3}{}_{1}4^{5}{}_{0}5^{4}{}_{1}1^{8}$

Discarding the remainders and the differences, we get our final answer as:

$$\frac{8}{31} = 0.25806451$$

Example 11: Convert $\frac{9}{51}$ to its decimal form.

A.F. of $\frac{9}{51} = \frac{9-1}{51-1} = \frac{8}{50} = \frac{0.8}{5}$

$\frac{0.8}{5} = 0.{}_{3}1^{8}{}_{3}7^{2}{}_{2}6^{3}{}_{3}4^{5}{}_{0}7^{2}{}_{2}0^{7}{}_{2}5^{4}$

$$\frac{9}{51} = 0.1764705$$

Case 4: When the denominator ends in 2, 3, 4, 6 or 7.

In these cases, the numerator and the denominator of the fraction will be suitable multiplied by a number to get the denominator to end in 9, 8 or 1.

For example, if we have the fraction where the denominator ends in 2, we can multiply the numerator and denominator by 4 to get a fraction that has denominator ending in 8 as shown below.

$$\frac{1}{22} = \frac{1}{22} \times \frac{4}{4} = \frac{4}{88}$$

We now have a fraction that has a denominator ending in 8. We already know the method for converting this fraction to decimal.

Similarly, for a denominator ending in 3, we multiply the numerator and denominator by 3.

$$\frac{3}{23} = \frac{3}{23} \times \frac{3}{3} = \frac{9}{69}$$

We now have a denominator ending in 9 and we can use the technique learnt earlier to convert this to decimal form.

Similarly,

$$\frac{5}{24} = \frac{5}{24} \times \frac{2}{2} = \frac{10}{48}$$

and

$$\frac{3}{26} = \frac{1}{26} \times \frac{3}{3} = \frac{3}{78}$$

and

$$\frac{5}{27} = \frac{5}{27} \times \frac{3}{3} = \frac{15}{81}$$

These fractions can now be converted to decimal form using the techniques learnt in previous sections.

Let us look at a few examples to understand this process.

Example 12: Convert $\frac{5}{32}$ to its decimal form.

Equivalent fraction:

$$\frac{5}{32}=\frac{5}{32}\times\frac{4}{4}=\frac{20}{128}$$

The equivalent fraction is $\frac{20}{128}$

And we now use the steps that we had used for a fraction whose denominator ends in 8.

Step 1: A.F. of $\frac{20}{128}=\frac{20}{128+2}=\frac{20}{130}=\frac{2}{13}$

The A.F. is $\frac{2}{13}$

Step 2: Start with the decimal point.

20 ÷ 13 gives 1 as quotient(Q1) and 7 as remainder.

To get the next GD we place the remainder before Q1 and add Q1 to it.

71 + 1 = 72. This is our GD for the next step.

$\frac{2}{13}=0._{7}1^{+1}$

Step 3: 72 ÷ 13 gives 5 as quotient (Q2) and 7 as remainder.

To get the next GD we place the remainder before Q2 and add Q2 to it.

75 + 5 = 80. This is our GD for the next step.

$\frac{2}{13}=0._{7}1^{+1}{}_{7}5^{+5}$

Step 4: 80 ÷ 13 gives 6 as quotient (Q3) and 2 as remainder.

To get the next GD we place the remainder before Q3 and add Q3 to it.

26 + 6 = 32. This is our GD for the next step.

$\frac{2}{13}=0._{7}1^{+1}{}_{7}5^{+5}{}_{2}6^{+6}$

Step 5: 32÷ 13 gives 2 as quotient (Q4) and 6 as remainder.

To get the next GD we place the remainder before Q4 and add Q4 to it.

62 + 2 = 64. This is our GD for the next step.

$\frac{2}{13}=0._{7}1^{+1}{}_{7}5^{+5}{}_{2}6^{+6}{}_{6}2^{+2}$

Step 6: 64 ÷ 13 gives 4 as quotient (Q5) and 12 as remainder.

To get the next GD we place the remainder before Q5 and add Q5 to it.

124 + 4 = 128. This is our GD for the next step.

$\frac{2}{13} = 0._{7}1^{+1}{}_{7}5^{+5}{}_{2}6^{+6}{}_{6}2^{+2}{}_{12}4^{+4}$

Step 7: 128 ÷ 13 gives 9 as quotient (Q5) and 11 as remainder.

To get the next GD we place the remainder before Q4 and add Q4 to it.

119 + 9 = 128. This is our GD for the next step.

$\frac{2}{13} = 0._{7}1^{+1}{}_{7}5^{+5}{}_{2}6^{+6}{}_{6}2^{+2}{}_{12}4^{+4}{}_{11}9^{+9}$

At this step it is clear that the GD is same as the previous step, so the quotient and remainder will keep repeating from this point on. Hence, we can write,

$\frac{2}{13} = 0._{7}1^{+1}{}_{7}5^{+5}{}_{2}6^{+6}{}_{6}2^{+2}{}_{12}4^{+4}{}_{11}9^{+9}{}_{11}9^{+9}{}_{11}9^{+9}{}_{11}9^{+9}{}_{11}9^{+9}{}_{11}9^{+9}$

Discarding the remainders and the additional numbers, we get,

$$\frac{5}{32} = 0.15624999999 = 0.15625$$

Example 13: Convert $\frac{3}{13}$ to its decimal form

Equivalent fraction of $\frac{3}{13} = \frac{3}{13} \times \frac{3}{3} = \frac{9}{39}$

Now the fraction is $\frac{9}{39}$. It has a denominator ending in 9.

Step 1: A.F. of $\frac{9}{39} = \frac{9}{39+1} = \frac{9}{40} = \frac{0.9}{4}$

Now applying the technique learnt in section 13.1, we get,

$\frac{0.9}{4} = 0._{1}2_{0}3_{3}0_{2}7_{3}6_{0}9_{1}2_{0}3$

Hence,

$\frac{3}{13} = 0.23076923$

Example 14: Convert $\frac{5}{17}$ to its decimal form

Equivalent fraction of $\frac{5}{17} = \frac{5}{17} \times \frac{7}{7} = \frac{35}{119}$

Now we have a fraction whose denominator ends in 9.

A.F. of $\frac{35}{119} = \frac{35}{119+1} = \frac{35}{120} = \frac{3.5}{12}$

Using the same technique as in example 13, we get

$\frac{3.5}{12} = 0._{11}2_{3}9_{3}3_{9}2_{8}7_{3}7_{1}3_{1}1$

Hence,

$$\frac{5}{17} = 0.29327731$$

It is clear from the above examples that we can convert any fraction to its decimal form in one single step eliminating the need for long division completely.

NOTE:

This method works only when the numerator is less than the denominator. But in case we have a fraction where the numerator is bigger than the denominator, we can convert it to a mixed number format and the whole part of the mixed number will form the digit before the decimal point.

Example: Find the decimal form of $\frac{49}{29}$.

Solution: We can write $\frac{49}{29} = 1\frac{20}{29}$.

Now our answer will begin with 1 followed by a decimal point.

Summary of steps

Case 1: When the denominator ends in 9.

Step 1: Find the Auxiliary Fraction of the given fraction by adding 1 to the denominator and then dividing both the numerator and denominator by 10.

Step 2: Start with the decimal point. Carry out a step by step division of the new numerator. At each stage the quotient forms the next digit of the answer and the remainder is placed before the quotient to form the Gross dividend for the next step.

The answer can be obtained to any degree of accuracy.

Case 2: When the denominator ends in 8

Step 1: Find the Auxiliary fraction of the given fraction by adding 2 to the denominator and then dividing the numerator and denominator by 10.

Step 2: Start the answer with the decimal point. Divide the numerator by the denominator. The quotient becomes the next digit of the answer and the remainder is placed before the quotient to form a number. The quotient is added to this number to get the Gross Dividend for the next step. This process continues and at every stage the quotient is added to the number obtained by placing the remainder before the quotient.

Case 3: When the denominator ends in 1.

Step 1: Find the Auxiliary fraction of the given fraction by subtracting 1 from the numerator and 1 from the denominator and then dividing both the numerator and denominator by 10.

Step 2: Start the answer with the decimal point. Divide the numerator by the denominator. The quotient forms the next digit of the answer and the remainder is placed before the difference of 9 and the quotient to get the Gross dividend for the next step. This process continues and at each stage we get the gross dividend by placing the remainder of the previous step before the difference of 9 and the quotient of the previous step.

Case 4: When the denominator ends in 2,3,4,6 or 7.

Step 1: Find an equivalent fraction of the given fraction by multiplying it with a suitable number so that the given fraction may end in 9, 8 or 1.

(Multiplication is summarized in table 13.1)

Step 2: Proceed with the steps given in the previous cases of denominators ending in 9, 8 or 1.

S.No.	Denominator ending with	Multiplication factor	Denominator will now end in	Example
1.	2	4	8	$\frac{5}{12} \times \frac{4}{4} = \frac{20}{48}$
2.	3	3	9	$\frac{4}{23} \times \frac{3}{3} = \frac{12}{69}$
		7	1	$\frac{4}{23} \times \frac{7}{7} = \frac{28}{161}$
3.	4	2	8	$\frac{5}{14} \times \frac{2}{2} = \frac{10}{28}$
4.	6	3	8	$\frac{5}{16} \times \frac{3}{3} = \frac{15}{48}$
5.	7	3	1	$\frac{5}{17} \times \frac{3}{3} = \frac{15}{51}$
		7	9	$\frac{5}{17} \times \frac{7}{7} = \frac{35}{119}$

Table 13.1

Try it yourself

Q1. Convert the following fractions to their decimal form. Give your answer up to 6 decimal places.

A.	(a) $\frac{1}{19}$	(b) $\frac{13}{29}$	(c) $\frac{15}{49}$	(d) $\frac{17}{69}$	(e) $\frac{27}{89}$
B.	(a) $\frac{3}{28}$	(b) $\frac{17}{38}$	(c) $\frac{21}{68}$	(d) $\frac{7}{88}$	(e) $\frac{37}{78}$
C.	(a) $\frac{11}{31}$	(b) $\frac{15}{21}$	(c) $\frac{18}{41}$	(d) $\frac{19}{81}$	(e) $\frac{43}{51}$
D.	(a) $\frac{7}{12}$	(b) $\frac{13}{33}$	(c) $\frac{15}{17}$	(d) $\frac{17}{36}$	(e) $\frac{27}{47}$

Chapter 14

Check – mate

In the previous chapters we have learnt many Vedic mathematics techniques which help us to do faster calculations. In some case, lengthy and complicated steps are eliminated leading to a simpler and easier calculation. It is a general tendency of students to check their answers after calculations. But it also happens that students are not able to trace their mistakes. For example, it may so happen that a student has written 3 + 3 = 9, it is likely that during rechecking also, the student will make the same mistake. In this chapter we will learn the technique of rechecking our calculations in such a manner that the mistakes can be traced and removed.

14.1 Digit Sum

Digit sum is the technique to convert a number to a single digit by repeatedly adding the digits of the number. For example, if the number is 976541, its digit sum or DS is calculated as

DS of 976541 = 9 + 7 + 6 + 5 + 4 + 1 = 32; 3 + 2 = 5.

DS of 976541 = 5

Note that the Digit sum will always be a single digit. We have to keep on adding the digits till we get a single digit answer.

A few more examples:

Number	Process	Digit Sum
581	5 + 8 + 1 = 14; 1 + 4 = 5	5
37564	3 + 7 + 5 + 6 + 4 = 25; 2 + 5 = 7	7
999871	9 + 9 + 9 + 8 + 7 + 1 = 43; 4+ 3 = 7	7
87543219	8 + 7 + 5 + 4 + 3 + 2 + 1 + 9 = 39; 3 +9 = 12; 1 + 2 = 3	3

This technique is not used to calculate any answer but to verify the answers that we have already calculated. Other than checking our calculations, this technique can also help during competitive exams where four choices are given, and the students have been asked to select one.

Short cut to find digit sum: We do not need to add all the digits to get the digit sum. We can eliminate all the nine's and all the digits that add up to 9. With the elimination of all nines and all the digits that add up to 9, we can calculate the digit sum much faster.

Example 1: Find the digit sum of 98378231

Normal method:

DS of 98378211 = 9 + 8 + 3 + 7 + 8 + 2 + 1 + 1 = 39; 3 + 9 = 12; 1 + 2 = 3

Short cut method:

We can eliminate 9, (8, 1) (7,2) (8,1)

~~9 8~~ 3 ~~7 8 2 1 1~~ = 3

As we can see that after eliminating all 9's and all the groups of digits adding up to 9, we are left with only digit 3. That is the digit sum.

Example 2: Find the digit sum of 979632013.

Normal method: 9 + 7 + 9 + 6 + 3 + 2 + 0 + 1 + 3 = 40; 4 + 0 = 4

Short cut method: ~~9 7 9 6 3 2~~ 0 1 3 = 1 + 3 = 4;

14.2 Procedure for checking answer

Case 1: Addition sum

Rule: The digit sum of the answer should be equal to the sum of digit sums of all the numbers used in the addition process.

That is, if

$a + b = c,$

then

$$DS(a) + DS(b) = DS(c)$$

Where DS denotes the digit sum of the respective numbers.

Example 3: Verify whether 7654219 added to 78993214 is equal to 86647433.

Eliminating the nines and the groups of digits which add up to 9, we get

Digit sum of the first number: 7654219 = 16; 1+ 6 = 7

Digit sum of the second number: 78993214 = 7

Sum of digit sums = 7 + 7 = 14; 1+4 = 5

Digit sum of the final answer: 86647433 = 41; 4 + 1 = 5

The digit sum of the final answer is same as the sum of digit sums of the numbers being added. Hence the answer is correct.

Example 4: Verify whether 92978413 added to 607542198 is equal to 700620611

DS of the first number: 92978413 = 7

DS of the second number: 607542198 = 6

Sum of the DS of both the numbers = 6 + 7 = 13; 1 + 3= 4

DS of the final answer: 700620611= 14; 1 + 4 = 5

The digit sum of the final answer is not equal to the sum of the digit sums of the individual numbers. Hence there is a mistake in addition.

Case 2: Subtraction

Rule: The digit sum of the smaller number as subtracted from the digit sum of the bigger number should be equal to the digit sum of the difference.

That is, if

$a - b = c,$

then

$$DS(a) - DS(b) = DS(c)$$

Example 5: Verify whether the following equation is correct:

$$95432168 - 435632 = 94996536$$

DS of the first number: 2

DS of the second number: 5

DS (first number) – DS (second number) = 2 – 5 = -3

In this case the difference of the digit sums is coming negative. We will add 9 to it to make it positive, since adding 9 or removing 9 does not change the digit sum of a number.

-3 + 9 = 6

DS of the difference: 6

Hence the answer is correct.

Example 6: Verify if the difference of 7982159 and 867024 is 7115145

DS of first number: 5

DS of second number: 18; 1+ 8 = 9.

If the digit sum comes as 9, we can take it as 0 also since 9 can be eliminated from the digit sum without affecting the answer

Difference of the digit sums = 5 – 0 = 5

DS of the difference= 6.

Hence the answer is incorrect.

Case 3: Multiplication

Rule: The product of the digit sums of the multiplier and the multiplicand should be equal to the digit sum of the product.

That is, if

$a \times b = c,$

then

$$DS(a) \times DS(b) = DS(c)$$

Example 7: Verify if the following equation is correct.

$$7312 \times 4351 = 31814612$$

DS of the multiplicand: 4

DS of the multiplier: 4

Product of the digit sums of multiplicand and multiplier= 4 x 4 = 16; 1 + 6 = 7

Digit sum of the product: 8

Hence the product is wrong. The equation is incorrect.

Example 8: Verify if the following equation is correct.

$$95643 \times 90979 = 8701504497$$

Digit sum of the multiplicand 95643 is 0 or we can take it as 9.

Digit sum of the multiplier: 90979 is 7

Their product: 7 x 9 = 63;

Digit sum of the product is 6 + 3 = 9

Digit sum of the product given in the equation: 8701504497 is 0 or 9.

Hence the equation is correct.

Case 4: Division

Rule: Use the formula

$$dividend = divisor \times quotient + remainder$$

That is, if

$a \div b \text{ gives } q \text{ as quotient and } r \text{ as remainder,}$

then

$$DS(a) = DS(b) \times DS(q) + DS(r)$$

Example 9: Verify if 2308682040 divided by 46532 gives 49614 as quotient and 43392 as remainder.

DS of divisor: 46532 is 2

DS of quotient: 49614 is 6

DS of remainder: 43392 is 3

DS of the dividend: 2308682040 is 6

Using the formula

$$DS(a) = DS(b) \times DS(q) + DS(r)$$

LHS = 6

RHS = 2 x 6 + 3 = 15; 1 + 5 = 6

LHS = RHS

Hence the calculation is correct.

Case 5: Square root

Rule: The digit sum of the square root as multiplied by itself should be same as the digit sum of the square.

That is, if

$\sqrt{a} = b,$

then

$$DS(a) = DS(b) \times DS(b)$$

Example 10: Verify if 29 is the square root of 841

DS of square root: 2+ 9 = 11; 1 + 1= 2

DS (sq. root) x DS (sq. root) = 2 x 2 = 4

DS of the number: 841 = 4

Hence it is correct.

Case 6: Cube root

Rule: The cube of the digit sum of the cube root should be equal to the digit sum of the cube.

That is, if

$\sqrt[3]{a} = b,$

then

$$DS(a) = DS(b) \times DS(b) \times DS(b)$$

Example 11: Verify if the cube root of 29791 is 31

DS of cube root 31 is 3 +1 = 4

DS (cube root) x DS (cube root) x DS (cube root) = 4 x 4 x 4= 64

DS of 64 is 6 + 4 = 10; 1+ 0 = 1

DS of the number: 29791 is also 1.

Hence it is correct.

14.3 Application

For students appearing for competitive examinations, this technique is very helpful. The students can check the alternatives given for the digit sum match with the question. This can help them eliminate some of the choices given.

However, this technique has a drawback. The digit sum method can only tell is if the answer is wrong or not. It cannot tell us with complete accuracy that the answer is correct or not.

This can be illustrated with the following example:

Example 12: Find the product of 9997 and 988

Assume that the student gets the answer as 9877036.

Checking the answer by the digit sum method, we get the digit sums are matching on both the sides. Hence, we can assume that the answer is correct. However, if the student had got the answer as 9877306, even then the digit sums would be same, although the answer is wrong.

Therefore, it is safe to say that if the digit sum DO NOT MATCH, then the answer is 100% wrong but if they match, then there is a possibility of the answer being right.

If the digit sums	**Then the answer is**
Match	Most likely correct
Do not match	Wrong

The summary of the rules discussed in this chapter is given in table 14.1.

Mathematical operation	Rule for checking
Addition	The digit sum of the answer should be equal to the sum of digit sums of all the numbers used in the addition process. if $a + b = c$, then $DS(a) + DS(b) = DS(c)$
Subtraction	Rule: The digit sum of the smaller number as subtracted from the digit sum of the bigger number should be equal to the digit sum of the difference. If $a - b = c$, then $DS(a) - DS(b) = DS(c)$

Multiplication	The product of the digit sums of the multiplier and the multiplicand should be equal to the digit sum of the product. If $a \times b = c$, then $DS(a) \times DS(b) = DS(c)$
Division	Use the formula $dividend = divisor \times quotient + remainder$ If $a \div b$ *gives q as quotient and r as remainder*, then $DS(a) = DS(b) \times DS(q) + DS(r)$
Square root	The digit sum of the square root as multiplied by itself should be same as the digit sum of the square. If $\sqrt{a} = b$, then $DS(a) = DS(b) \times DS(b)$
Cube root	The cube of the digit sum of the cube root should be equal to the digit sum of the cube. If $\sqrt[3]{a} = b$, then $DS(a) = DS(b) \times DS(b) \times DS(b)$

Table 14.1.

Try it yourself

Q1. Verify whether the following answers are correct or incorrect without actual calculation

a) 3125 + 45693 + 76542 + 13245 = 138605
b) 4521098 + 6347621 + 9097652 = 19966471
c) 6983451 – 78453 = 6904898
d) 6582170 – 998979 = 5583291
e) 0.4565 x 0.653 = 0.2980945
f) 95231 x 9992 = 951548152
g) 598650 divided by 125 gives 4789 as quotient and 25 as remainder
h) 9000 divided by 120 gives 75 as quotient
i) Square root of 783225 is 885
j) Cube root of 30664297 is 303

ANSWERS

Chapter 1

Q1.	a) 357	b) 274	c) 155	d) 8272	e) 3141
	f) 800	g) 1902	h) 29800	i) 799	j) 2800
Q2.	a) 76.58	b) 41.28	c) 761.877	d) 5478.91	e) 346.785
	f) 8.7653	g) 71690.9	h) 8798.95	i) 36.745	j) 3596.72

Chapter 2

Q1.	a) 2178	b) 1053	c) 4372	d) 9803	e) 72650
	f) 3325	g) 23795	h) 11008	i) 1232	j) 2589

Chapter 3

Q1.	a) 3564	b) 7722	c) 6831	d) 4752	e) 2178
	f) 344655	g) 785214	h) 253746	i) 860139	j) 466533
	k) 23447655	l) 56734326			
Q2.	a) 66933	b) 4649535	c) 3209679	d) 96903	e) 319968
	f) 5639436	g) 2209779	h) 85914	i) 939906	j) 36499635
Q3.	a) 4562433	b) 46332	c) 368379	d) 49203	e) 62568
	f) 5158836	g) 4216779	h) 87714	i) 48906	j) 28215

Chapter 4

Q1.	a) 7073	b) 64592	c) 9636	d) 717552	e) 479631
	f) 38295	g) 284937	h) 363858	i) 8503594	j) 8500261
	k) 10923352	l) 726594			
Q2.	a) 4092	b) 5700	c) 4668	d) 8148	e) 6504
	f) 27744	g) 104772	h) 112704	i) 98112	j) 83776
	k) 4204472	l) 8252152			
Q3.	a) 5967	b) 11475	c) 23312	d) 4806	e) 40672
	f) 4212	g) 11648	h) 17841	i) 12971	j) 5852
	k) 13756	l) 9882			

Chapter 5

Q1.	a) 7626	b) 9801	c) 8554	d) 986045	e) 990024
	f) 984055	g) 967200	h) 99880035	i) 99750156	j) 99660253
Q2.	a) 11340	b) 12426	c) 11322	d) 1009020	e) 1022112
	f) 100210104	g) 100150044			
Q3.	a) 8091	b) 9975	c) 10137	d) 997985	e) 994896
	f) 1001992	g) 998868			
Q4.	a) 86217	b) 93186	c)82585	d)910364	e) 103064
	f)107460	g) 93718	h) 97388		

Chapter 6

Q1.	a) 2730	b) 3298	c) 2352	d) 6478	e) 2465
Q2.	a) 234255	b) 176276	c) 221288	d)129492	e) 237168
Q3.	a) 16374708	b) 24429629	c) 28092489	d) 29825092	e)20403834

Chapter 7

Q1.	a) Q=81; R=3	b) Q=78; R=7	c) Q=44; R=7		
Q2.	a) Q=21; R=55	b) Q=35; R=81	c) Q=66; R=81	d)Q=128; R=53	e) Q=458; R=6
Q3.	a) Q=34; R=546	b) Q=46; R=219	c) Q=21; R=60		

Chapter 8

Q1.	a) Q=33; R=6	b) Q=26; R=9	c) Q=594; R=7	d) Q=341; R=7	e) Q=231; R=10
Q2.	a) Q=13; R=96	b) Q=29; R=69	c) Q=149; R=2	d)Q=240; R=21	e) Q=223; R=36
Q3.	a) Q=44; R=803	b) Q=19; R=283	c) Q=116; R=385	d) Q=283; R=657	e) Q=132; R=782

Chapter 9

Q1.	a) Q=60; R=36	b) Q=160; R=32	c) Q=1946; R=13	d) Q=70; R=21	e) Q=244; R=13
	f) Q=1432; R=12	g) Q=2461; R=15	h) Q=197; R=19	i) Q=3078; R=14	j) Q=1230; R=61
Q2.	a) Q=1613; R=30	b) Q=592; R=54	c) Q=3390; R=122	d)Q=230; R=312	e) Q=2319; R=222

Chapter 10

A	a) 3025	b) 4225	c) 7225	d) 9025	e) 13225
B	a) 1764	b) 2809	c) 3364	d) 3481	e) 2304
	f) 2116	g) 3136	h) 2601	i) 2401	j) 2209
C	a) 11236	b) 11881	c) 10816	d) 11664	e) 12544
D	a) 8281	b) 8649	c) 7744	d) 8464	e) 8836
E	a) 1008016	b) 1018081	c) 1022121	d) 1006009	e) 1024144
F	a) 996004	b) 990025	c) 994009	d) 976144	e) 982081
G	a) 6084	b) 4225	c) 271441	d) 228484	e) 454276
	f) 16900321	g) 9474084	h) 25030009	i) 65464281	j) 85266756

Chapter 11

Q1.	a) 99	b) 75	c) 44	d) 59	e) 37
	f) 58	g) 79	h) 84	i) 56	j) 92
Q2.	a) 213.62	b) 280.45	c) 286.95	d) 314.23	e) 180.15
	f) 422.67	g) 753.22	h) 589.62	i) 899.96	j) 675.51

Chapter 12

Q1.	a) 38	b) 49	c) 73	d) 67	e) 59
	f) 88	g) 51	h) 34		
Q2.	a) 75	b) 72	c) 87	d) 82	e) 36
	f) 99	g) 69	h) 78		
Q3.	a) 438976	b) 636056	c) 42875	d) 753571	e) 103823
Q4.	a) 857375	b) 1225043	c) 704969	d) 658503	e) 1295029
	f) 1367631	g) 1006012008	h) 99401192	i) 967361669	j) 1012048064

Chapter 13

A	a) 0.052631	b) 0.448275	c) 0.306122	d) 0.246376	e) 0.303370
B	a) 0.107142	b) 0.447368	c) 0.308823	d) 0.079545	e) 0.474358
C	a) 0.354838	b) 0.714285	c) 0.439024	d) 0.234567	e) 0.843137
D	a) 0.583333	b) 0.393939	c) 0.882352	d) 0.472222	e) 0.574468

Chapter 14

Q1.	a) Yes	b) No	c) No	d) No	e) Yes
	f) Yes	g) Yes	h) Yes	i) Yes	j) No